AF443049

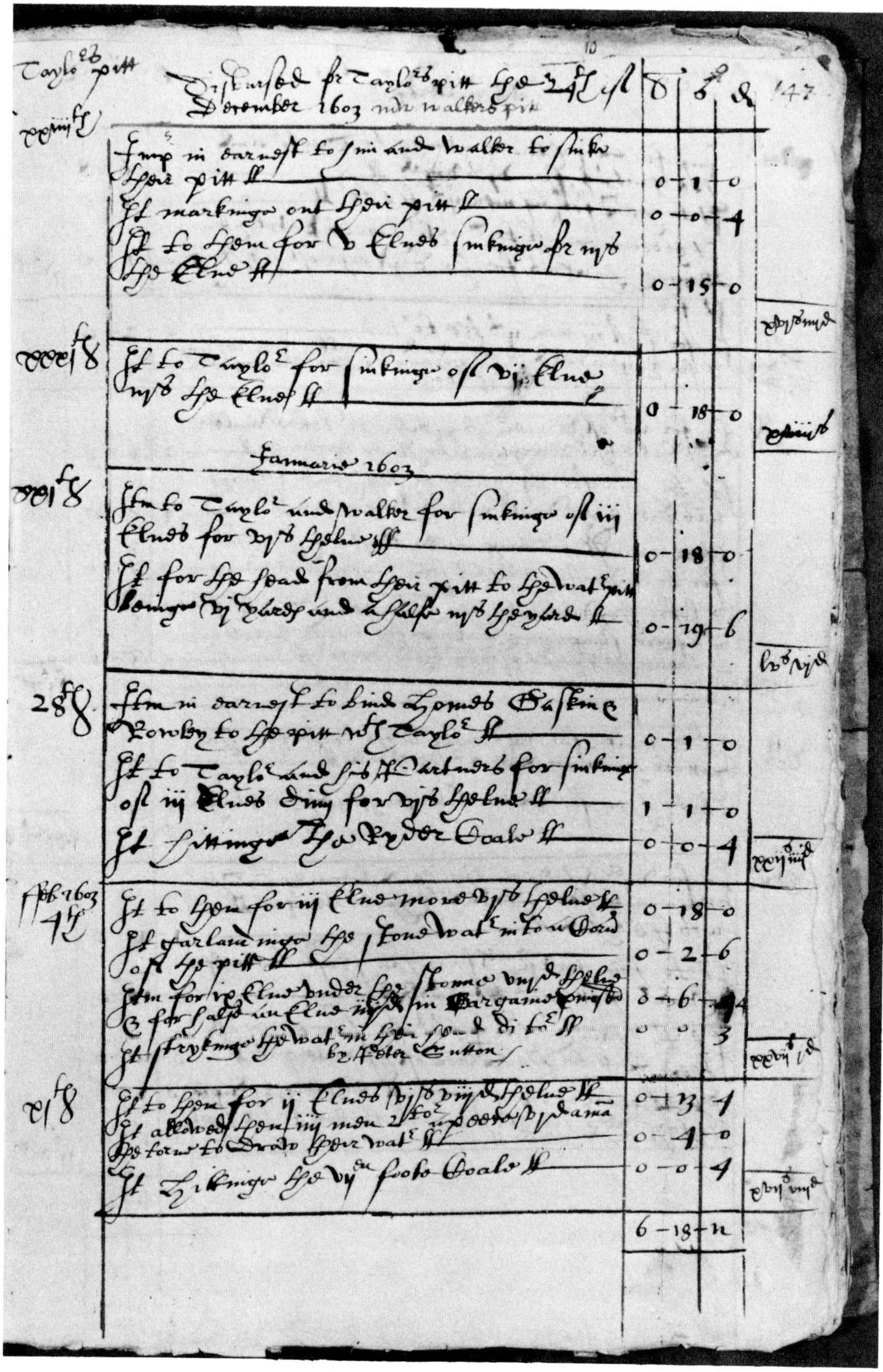

Taylo^{rs} pitt

Disbursed for Taylo^{rs} pitt the 24 of
December 1603 nder walkers pitt   l   s   d   147

Imp^r in earnest to Smi and walker to shike
theire pitt ℔ _______________________________ 0 — 1 — 0

It markinge out theire pitt ℔ _______________ 0 — 0 — 4

It to them for 2 Clues sinkinge in to
the Clue ℔ _________________________________ 0 — 15 — 0

It to Taylo^r for sinkinge of vij Clue
in to the Clue ℔ __________________________ 0 — 18 — 0

Januarie 1603

It to Taylo^r and walker for sinkinge of iiij
Clues for vjd the Clue ℔ __________________ 0 — 18 — 0

It for the road from theire pitt to the wat^r pitt
bonge vij yardes and a halfe in the yard ℔ __ 0 — 19 — 6

It^m in earnest to bind Thomas Gaskin
Rowley to the pitt wth Taylo^r ℔ ___________ 0 — 1 — 0

It to Taylo^r and his ij arbeders for sinkinge
of iij Clues ding for vjd the Clue ℔ _________ 1 — 1 — 0

It hittinge the bynder Coale ℔ ______________ 0 — 0 — 4

It to them for iij Clue more vjd the Clue ℔ ___ 0 — 18 — 0

It for landinge the stone wat^r in to a bord
of the pitt ℔ _____________________________ 0 — 2 — 6

It^m for ix Clue under the stonne vij d the Clue
& for halfe in Clue in th in bargaine vj^d ____ 0 — 6 — [ ]
                            0 — [ ] — 3
It stykinge the wat^r in theire pitt & di th ℔
by Peter Sutton

It to them for ij Clues vij^d viij^d the Clue ℔ _ 0 — 13 — 1

It allowed them iiij men 2 s 2d in earnest by daua
the for w^r to draw theire wat^r ℔ ______________ 0 — 4 — 0

It hittinge the vj de bore Coale ℔ ____________ 0 — 0 — 4

                           6 — 18 — 11

Fig. 1

DUGDALE SOCIETY OCCASIONAL PAPERS
No. 26

A Warwickshire Colliery in the Seventeenth Century

BY

E. G. GRANT

LEEDS
Printed for the DUGDALE SOCIETY
by W. S. MANEY & SON LIMITED
1978

*Printed in England
by W. S. Maney & Son Limited
Hudson Road Leeds LS9 7DL*

CONTENTS

The Dugdale Society wishes to record its thanks to
the National Coal Board South Midlands Area for
financial assistance towards the publication of this
Occasional Paper.

A WARWICKSHIRE COLLIERY IN THE SEVENTEENTH CENTURY

THE Warwickshire coalfield is one of several small coalfields that together comprise the Midland coalfield lying in the counties of Stafford, Warwick, Leicester and Derby. It consists of a narrow exposed coalfield (i.e. the area where the coal seams reach the surface) occupying a long tract of countryside from Hawkesbury (just north of Coventry) northwards through Bedworth and Stockingford to Polesworth and then sweeping southwestwards through Wilnecote to Dosthill. South and west of the coalfield lies an extensive and somewhat deeply buried concealed coalfield that was not exploited, or indeed discovered, until the nineteenth century.[1]

Sporadic working of coal, mainly by surface quarrying, took place during Roman and medieval periods but were of little consequence in determining later patterns. It was not until the second half of the sixteenth century that any real progress was made with the advent of shaft mines, primitive drainage equipment, and a more capitalistic approach to the industry. The emergence of the Warwickshire coalfield during this period was part of a national trend that has been investigated in great detail by J. U. Nef. Nef went as far as to proclaim that the period from 1550–1700 was the most important period of growth in the British coal industry and though some of his figures for the output of coal during the period are now considered to be exaggerations there is little doubt 'that there appears to have been something like a revolution in the use of fuel in the period'.[2]

There are several reasons for this sudden explosion, the most important and most obvious one being a remarkable development in the demand for coal. Domestic and industrial consumption increased not only *per capita*, but as time went on to include a growing population of more and more people, many of whom had never seen or heard of coal previously, but soon became totally dependent on it. This in turn was due to a national crisis in the supply of timber, caused by excessive and frequently reckless exploitation of natural timber for firewood, charcoal burning, industrial heating, shipbuilding and domestic house construction. The price of firewood increased several times as the supply dwindled and many people found themselves being, as the Bishop of London lamented, 'driven to burn seacoles',[3] often against their judgement as coal burning was considered injurious to health. The smokiness of coal compared with dry firewood encouraged people to install chimneys in their house where previously wood smoke had found any convenient exit.

An exactly similar growth pattern was manifested by industrial demand for coal. First, coal was substituted for firewood and charcoal in a whole range of industries, including glass-making, salt-making, lime-burning, brick- and tile-making, blacksmithing and dyeing. Secondly, these industries, and several new ones such as clay-pipe making, experienced an early industrial revolution, so that output was much greater than before, which encouraged an even greater demand for coal. By the end of the seventeenth century industry was using one-third of the total output of coal (estimated by Nef at 3 million tons, almost certainly an exaggeration[4]), but it was effort well spent, as by that date the timber crisis had abated and woods and forests were beginning to appear again in the English landscape.[5]

This vastly increased demand for coal was met by a vastly increased supply that was in turn only made possible by major advances in mining technology and a near revolution in the patterns of wealth and land-holding. The European renaissance in science and technology had nurtured considerable progress in the methods of sinking, operating and draining mines, processes that are well described and illustrated in Agricola's *De Re Metallica* of 1556. In England the dissolution of the monasteries had allowed land to be redistributed from church to crown and then to private ownership, creating a new class of lay landowners. These landowners were usually keen to exploit any coal that lay under their estates frequently by finding as partners a new breed of men known as 'adventurers'. These spirited entrepreneurs were the link between the technical and the financial sides of the industry, and often recklessly poured their own and anyone else's capital into the rapidly proliferating coal mines.

The Warwickshire coalfield became established as one of the important coalfields of the Midlands, and landowners, merchants and adventurers invested considerable quantities of money in the mines. Production increased quite spectacularly from (at most) 2,000–3,000 tons per annum in 1550, to about 40,000 tons in A.D. 1700.[6] Nef calculated that the total output was nearer 70,000 tons, but this has to be seen as potential output rather than a constantly maintained figure. If all the collieries were in simultaneous production the figure would have been reached and exceeded, as Warwickshire had some of the largest collieries in the country. There were, however, few if any years in which they all operated at full capacity, due to technical and financial problems inherent in the period.

The total amount of coal raised in Warwickshire was not large by the standards of other fields (Nef estimated 100,000 tons per annum for Staffordshire, and over 1 million tons for the Northumberland and Durham field) but the small size of the coalfield and the fact that the activity was largely concentrated in the southern part of the area, meant that the sudden increase in the amount and scale of mining activity there was manifested as an industrial revolution. Almost the whole of the exposed coalfield from Nuneaton southwards was covered by leases made for mines[7] of coal in Nuneaton, Chilvers Coton (Griff), Bedworth, Foleshill and Sowe.[8] Coal was also being worked in the northern part of the coalfield, for example at Dordon, but less continuously (in a spatial sense) and less well-known from a documentary point of view. Some of the most important personages in the emergence of the British coal industry during this important period were attracted to the Warwickshire coalfield. Sir Francis Willoughby and Nicholas Beaumont, coal masters from Wollaton in Nottinghamshire and Coleorton in Leicestershire respectively were operating the important and productive Bedworth colliery in the 1570s.[9] Willoughby also leased mines at Foleshill and Sowe in the 1580s,[10] in partnership with Sir Clement Fisher of nearby Packington Hall who probably put up much of the capital required.

By the end of the sixteenth century it was the Beaumont family who dominated the scene. Two of Nicholas Beaumont's sons, Sir Thomas and Huntingdon Beaumont, had obtained, by leasing or buying out others, control of all the mines in the parishes of Bedworth, Foleshill and Sowe.[11] Huntingdon Beaumont was a brilliant and colourful adventurer who dabbled in mines in most of the Midland coalfields as well as an abortive attempt in the Northumberland and Durham coalfield.[12] The Willoughbys and Beaumonts have to be regarded as the founding fathers of the emergent Warwickshire coal industry and their enthusiasm and national standing was in no small way responsible for publicizing the existence and relative importance of the Warwickshire coalfield. By the 1630s, for example, the colliery at Bedworth, which they had first operated, was producing 20,000–30,000 tons per annum, the largest colliery in the county and one of the largest in the country.[13]

There was, however, considerable variation in the size of collieries at this time, and few were as large as the Bedworth undertaking. Collieries producing 5,000–20,000 tons per annum were more usual and can be described as medium sized collieries. Small collieries producing less than 5,000 tons were also frequently opened up but tend to survive less well in the documentary record. Even so, changes in the scale of colliery operation tended only to affect the capital input rather than the operational process.

Immediately north of Bedworth lies the area known as Griff, comprising the southern part of the parish of Chilvers Coton. Griff had been worked for coal during the medieval period[14] and mines were re-established there by Geoffrey Foxe during the period of activity before the end of the sixteenth century.[15] A medium sized colliery which operated at Griff from 1603–5 has been chosen for detailed analysis as an example of the kind of enterprise that could be found in Warwickshire during the seventeenth and last part of the sixteenth centuries. It was a somewhat short-lived enterprise though its closure due to insufficient profits was frequently repeated for other undertakings up and down the exposed coalfield. A further factor influencing the choice of this colliery for detailed treatment is that a set of accounts survive for part of its operational period, and these form the basis for this paper. These accounts are described below, their major drawback being that they were prepared for an accountant's eyes, and thus emphasize expenditure, income and cash balances rather than production or locational attributes. Lawrence Stone lamented the lack of detailed investigation of early colliery accounts, as it is only by their examination that a basic corpus of knowledge on costs, production, profits and labour can be established. Stone himself used the accounts of the Earl of Shrewsbury's Colliery at Sheffield Park, Yorks., to illustrate the kind of information that could be obtained[16] and the present study can be seen as a contribution to establishing a broader base for the elucidation of a typical Midland colliery of the seventeenth century.

1. *The Griff Account Books*

The documentary evidence for the colliery, which belonged to the Newdegates of Arbury, is provided by the survival of 'The booke of accomptes for the Slate Cole delphe', and 'The Booke of accomptes for the Seven Foote delphe'.[17] Physically these accounts occupy one notebook of over 100 sheets of paper (226 sides), bound in vellum. The entries were made by William Henshaw, clerk of the coalworks, in a densely written secretary hand which, though legible, is not easy to read (Fig. 1). Spelling is erratic, and quantities and moneys are frequently given in Roman numerals. The two main divisions of the book into the Slate Coal pits and the Seven Foot pits are further divided into sections dealing with sinking and operating, drainage, sales and production, charter, wages, timber, blacksmith's charges, etc. There were originally blank pages between each section, but they were used for some calculations and entries almost 100 years later. In the part of the book containing the Seven Foot accounts, there was insufficient space for entering the sales and production figures and the charter costs, so these were written at the end of the Slate Coal accounts, thus making the entries rather difficult to follow.

Almost at the end of the notebook the final calculations and cash balances were made, including some entries of no relevance to coal mining, one of which recorded £25 disbursed 'for the behoofe of S^r Jo Newdigate wch he then did leand to the King upon a privie seal' (App. p. 225). The final balances were merely cash balances so that the clerk could account for all sums advanced to him by William Whytehall (who acted as chief accountant for the coal works) and all cash sales for coal, miners' wages etc. Nowhere was there any attempt at calculating profits or losses, nor any comments on the progress of the colliery.

The special arrangement between Sir John Newdegate and Sir Richard Leveson and Francis Fitton is evidenced by the fact that he (or his son) signed some of the final balances, and, after the colliery closed, the account book became the possession of Sir John, to survive among the Newdegate MSS to this day. The later insertions in the blank pages are also interesting. They were made in A.D. 1700 by Sir Richard Newdegate, Sir John Newdegate's grandson, who was also a major coal-master in Warwickshire. He made an index of his interpolations (as well as an index for the whole book), listing calculations for his sough, colliers' names, computations of profit and loss, and many other details for the colliery he operated in about A.D. 1700. Unfortunately most of his pages were later ripped out, only eight surviving from what would have been an immensely useful catalogue of ideas and requirements for setting up a colliery.

2. *Ownership and Location*

The Newdegates were fairly recent arrivals in Warwickshire when their first pits were recorded in 1603. John Newdegate exchanged his estate at Harefield, near Uxbridge in Middlesex, in 1585 for Arbury Priory and manor in the parish of Chilvers Coton, owned by Sir Edmund Anderson, Chief Justice of the Common Pleas.[18] No prior indication of any mining interest by the Newdegates is given before their entry on the scene in 1603, but they were probably influenced by two main factors. First, the activities of the mining adventurers had reached its peak by 1603, and Arbury lay close to the scene of their industry. Sir Thomas Beaumont was working the mines of Bedworth, immediately to the south of the Arbury estate, while on a part of Griff belonging to the

Gifford family lay an ill-fated colliery operated by Geoffrey Foxe. The second factor was that the outcrop of the coal seams ran right along Griff estate, at least part of which had just been absorbed into the Arbury estates. Griff manor originated in the twelfth century when Ralph Sudeley divided his manor of Chilvers Coton into three, one part to endow Arbury Priory which he had founded, another to the Knights Templars, and the third part — Griff — being retained for his own use.[19]

Griff manor seems to have occupied about one square mile in the south-east of Chilvers Coton parish, between the boundary of Bedworth on the south and Griff brook on the north, beyond which was the manor of Temple (see Fig. 2). The site of Sudeley Castle and Griff hamlet occupied an approximately central position in the manor. From the Sudeleys the manor passed to the Boteler family and in 1561 was sold to the Giffords, a Staffordshire family, who also purchased Temple Manor to the north. The history of Griff becomes obscure about 1600 when the manor appears to have been split into northern and southern estates — Sudeley Castle was the dividing line (see Fig. 2) — and it was the southern estate that came into the possession of the Newdegates.[20]

The emergence of the Newdegates as coal-masters introduced a new type of person into the coalfield, the local landed gentleman. Landed gentry had already appeared in the coalfield, but they were not local. Sir Francis Willoughby was the most important of these and, though he owned land in Warwickshire, he could not be considered a local, and was much closer to the adventurer in spirit. Sir John Newdegate, the first mining member of the family, was a representative of the essentially conservative land-owning class to whom working the minerals on their land was considered as another form of estate income, similar to raising cattle or timber. There is plenty of evidence that this was the continued outlook of the Newdegates with respect to their mineral rights.[21] There were times when mining waxed important and additional land would be rented for the purpose, but equally, if coal did not fit into the estate scheme at any one time, the mines were leased out or abandoned. The history of the Newdegate mines is a continuous alternation from direct working to leasing, and back to direct working whenever an incumbent Newdegate became again interested in trying to profit from the mines.

Sir John Newdegate commenced working coal at Griff in 1603, but only a few months after commencing let the mines and all mineral rights of coal on the estate to Sir Richard Leveson and Francis Fitton.[22] Though the lease was properly drawn up in legal terms, it was in effect a special arrangement among friends and relatives, and in that respect is not strictly comparable with other leases of the coalmines. The person-

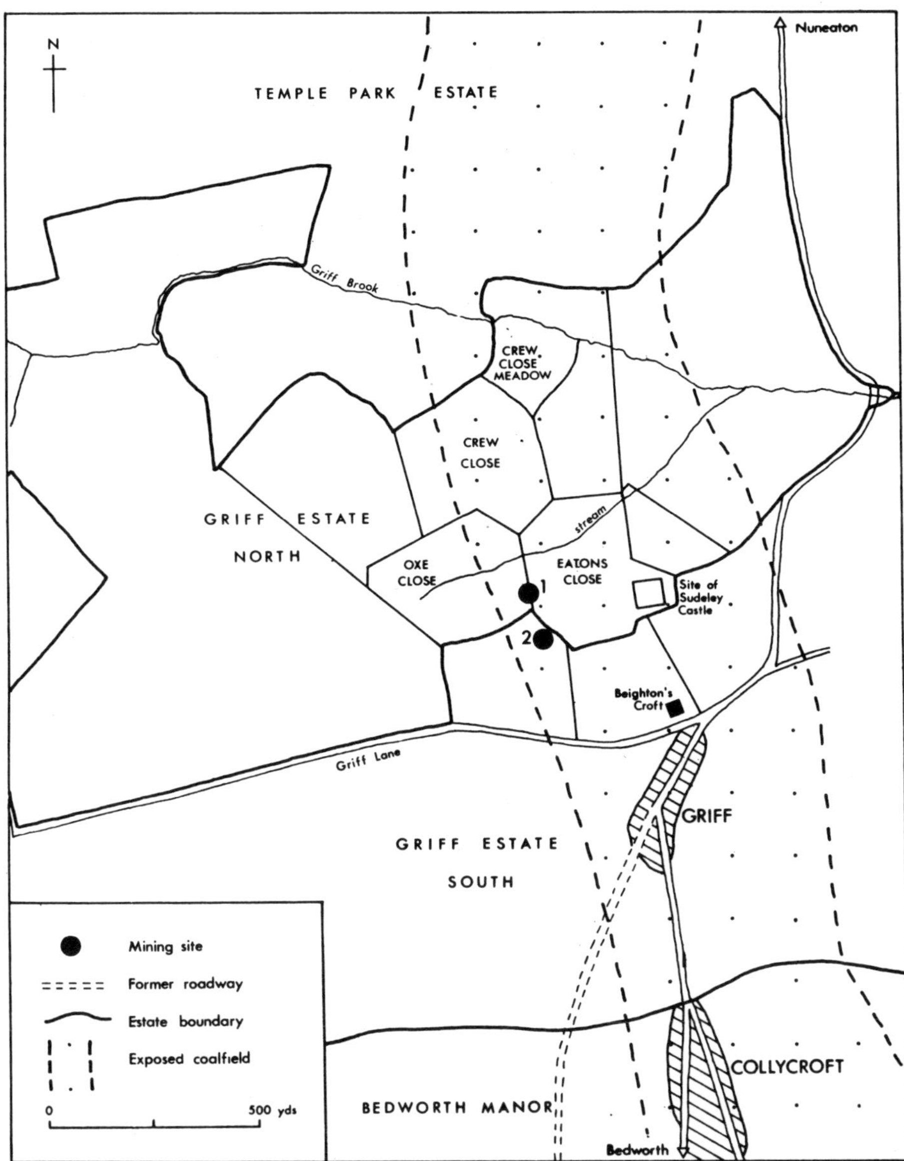

Fig. 2 <u>Seventeenth Century Mining Sites at Griff</u>. Sites are only marked where the actual locations are known. The field boundaries in Griff Estate North are taken from a late 17th. cent. map of that estate. (W.C.R.O. CR319/60.)
1. Site worked by Foxe, Thomas Beaumont, and Collins and Potter.
2. Site of Griff Colliery, 1603-5.

age and relationship of Sir Richard Leveson is unknown and he seems to have played no part in the subsequent operation of the colliery. Francis Fitton was probably Sir John Newdegate's brother-in-law. Included in the leasor side with Sir John was William Whytehall who lived at Arbury, and was the uncle of Francis Fitton and Lady Newdegate. The terms of the lease were remarkably generous; in return for holding all the mines sunk by 1603, and all others that might be found, along with total freedom of access and operation, the lessees were to pay 'so many of the best coles as Sir John Newdegate shall yerely fetch to burn only at his house called Arbury and not elsewhere'. The other main agreement was that Sir John would not 'intermeddle' with the operation of the mines without the assent of Leveson or Fitton in writing.

The only interpretation that can be put on this curious arrangement is that Sir John agreed or was persuaded that the others could run the colliery better than himself, and he was happy to allow them, so long as he received sufficient coal to keep his home fires burning. From 6 August 1603 to 27 October 1604, 109 loads of coal (approximately 150 tons) were sent to Arbury for that purpose, though it only amounted to 3% of the output of the colliery during the period, an exceptionally small rent.

The accounts do not give an exact location for the pits, other than that they were at Griff. However, a separate plan of the colliery drawn up in 1604 provides some evidence, as Beighton's Croft is shown lying to the south and Sir Thomas Beaumont's land to the north (see Fig. 3 and accompanying note). Beighton's Croft (now the area around Griff House) is a well-known location (Fig. 2), on the north side of Griff hamlet, so the site of the colliery was immediately to the north of that, near the site of Sudeley Castle and right on the boundary of the recently subdivided Griff estate. The reference to Sir Thomas Beaumont having an interest in the land immediately to the north is interesting,[23] as that lay in the northern Griff estate, still in the possession of the Giffords, who had leased Geoffrey Foxe's pits to Sir Thomas. In fact, the two collieries — Newdegate's and Beaumont's — were contiguous, with the new estate division running between them. There are frequent references in the 1603–5 accounts to Beaumont's colliery, and on one occasion his miners accidentally broke through to the Newdegate pits (App. p. 151). Beaumont's drainage equipment also unintentionally helped to drain the Newdegate pits, and he unsuccessfully tried to draw up a joint drainage scheme with Francis Fitton in 1604.[24]

We do not know how long Sir John Newdegate worked Griff colliery before leasing it to Leveson and Fitton, but when these operators took it over, five pits were already sunk and in production, and indeed were quickly exhausted, hence the need for sinking new pits on the outcrop of the Seven Foot Coal.

3. *Sinking*

As explained above, the Griff account book is in two parts, one for the existing Slate Coal pits from 30 July to 5 November 1603, and the other covering the new Seven Foot Coal pits from the week ending 19 November 1603 to 27 October 1604. At the end of the Seven Foot accounts a final page of summations gives bare details of the continuation of the Seven Foot pits until 27 April 1605, and one or two isolated expenses were entered until the final balances were made on 20 January 1607 (App. p. 225).

The Slate Coal pits were already in production when their account opens, but the Seven Foot pits were started from scratch, and provide an excellent description of the opening-up of a colliery.[25] Even before the Slate Coal pits were finally exhausted, trials were being made for the Seven Foot Coal. In October 1603, a miner called Wootton whose Slate Coal pit was finished sunk two trial pits in 'Smythes Close' (location uncertain), but with no success, and after a third pit had been tried in the following February, the pits were filled in.[26] Smith's Close was probably too far east beyond the outcrop of the Seven Foot Coal, but a successful strike was soon made, and the new colliery was commenced. The layout that was being planned consisted of a row of three main pits, 86–88 feet deep, paralleled by a row of three shallow pits, plus another shallow pit, an air pit and a water pit (see Fig. 4). The three main pits were sunk, 'by the companies of iii pits men viz Walker, Evans & Taylor being the head of every severall pitt by their names so severallie called',[27] that is, each pit was known by the name of the headman of the group of miners employed there. Each company, or partners as they were more often called (usually four to six men), were paid 10*d.* for every three-quarters[28] of coal they raised, but if they were employed in other jobs such as drawing water or heading galleries, they were paid a standard wage. It was up to the headman to share out the 'chalter' (i.e. charter) payment of 10*d.* a three-quarters, and it was obviously in his interests to have as efficient a team as possible, composed of hewers, hauliers and banksmen (who raised the coal and stacked it on his company's bank). If a team managed to raise 20 three-quarters in a day, another payment of 4*d.* for that day was earned (see App. p. 116 for chalter payment and p. 154 for the productivity allowances).

This method of labour organization appears to have been unique to Warwickshire. Charter systems were not unknown elsewhere but they

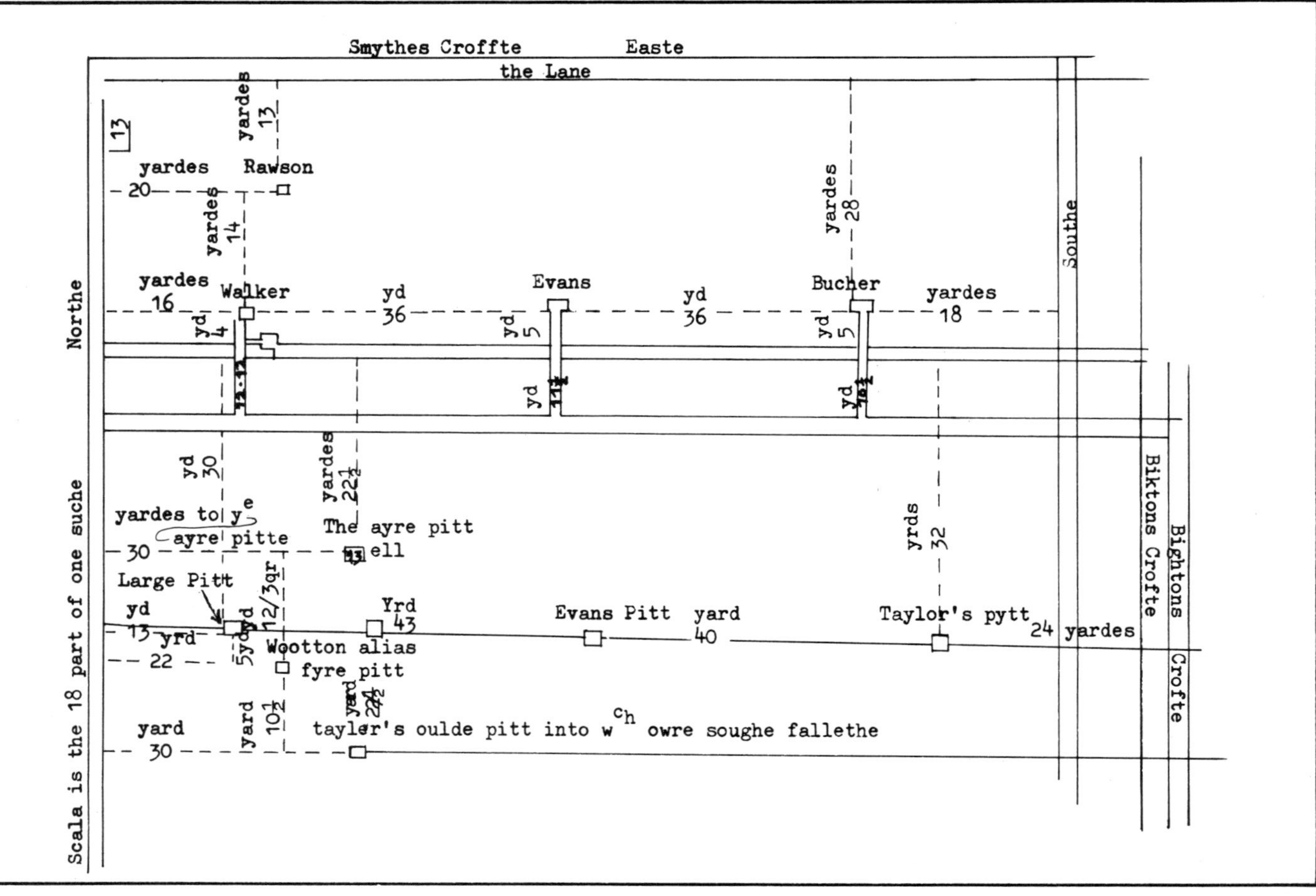

Fig. 3   Plan of Griff Colliery, 1604.

were usually a looser system of organization. For example, in 1637 the Bishop of Chester made a contract with four colliers to work his pits at Farnworth in Lancashire in return for 8*d*. a quarter (approximately 2*s*. per ton),[29] their higher remuneration no doubt being related to their having to finance all the underground working, unlike the Griff miners who seem to have had the best of both worlds, that is, payment by results when the going was good and payment of a fixed wage when other jobs were necessary.

The first pit to be sunk, other than the water pit for drainage, was Walker's pit, though when it was first commenced Taylor was the

Interpretation of the Plan of Griff Colliery, 1604 (Fig. 3)

The original plan (W.C.R.O., CR136/C.489) measures approximately 12″ × 18″ and on the back bears the inscription:

'xi[th] March 1603 (i.e. 1604)
A plotte of shewe howe all the pyttes are placed and sytuated; with theyre distances eche from the other in the collecrofte, upon the 7 foote cole/'

A trial version on the back of the plan contains the additional information that on the north side a lane parted the colliery from Sir Thomas Beaumont's demise.

Unfortunately, the plan is not quite clear or complete enough to be totally correlative with the entries concerning the layout of the colliery given in the 1603–4 account book. The date of the plan is interesting, because in the week ending 10 March the basset pits were commenced, and the plan may well have been drawn up to indicate the position of these. The shafts on the east side of the diagram marked Rawson, Walker, Evans and Butcher are considered to be the basset pits and the row of pits to the west to be the main coal pits. The distance between the two rows of pits was about 35 yards which agrees with the expected distance calculated from the depths of the pits and the dip of the coal. (The coal seam had risen *c.* 35 feet between the main pits and the basset pits, and with a dip of one in three, was equivalent to a horizontal distance of 3 × 35 feet = 35 yards.)

The indication of the parallel lines immediately to the west of the basset pits — and the short ones at right angles — is not clear. They may refer to underground shafts and heads, but all the other detail on the map is surface phenomena and all the distances were measured on the surface. They could be interpreted as paths or boundaries — known boundaries and lanes are indicated in the same format — which were used as fixed positions for the measurements to lay out the basset pits which are marked in a very regular fashion.

The row of main pits also provides interpretative problems, as Evans's pit and Taylor's pit are named, but not Walker's. Of the ancillary pits, the air pit, Taylor's old pit and Wootton's pit (the last two were survivals from the Slate Coal workings) are clearly indicated, but not the water pit. The plan is thus lacking in positions for Walker's pit and the water pit, but on the other hand there is an un-named pit and another simply called the Large Pit. I would suggest that the Large Pit is Walker's pit (this puts it in the right position *vis-à-vis* Evans's and Taylor's/Butcher's pits) and the un-named one the water pit. A simplified plan of the layout and a typical cross-section are shown in Fig. 4.

headman (see App. pp. 147–55 for a complete account of Walker's pit).
In the week ending 24 December 1603 Taylor and Walker were given 1*s*.
'in earnest . . . to sinke their pitt' and 4*d*. for 'markinge out their pitt'.
The giving of earnest money (in reality a tip) was widespread at this
time. In the first week they sunk 5 ells at 3*s*. an ell, and 6 ells the
following week. After a three week break at New Year they resumed
work by sinking 3 ells at the increased rate of 6*s*. an ell at which point
(52½ feet down) they started a head[30] to the water pit, 6½ yards long.

So far Taylor and Walker were working by themselves but at the end
of January 1*s*. was paid to 'bind Thomas, Gaskin and Rowley to the pitt
wth Taylor', and they were then known as 'Taylor and his Partners'.[31]
Another 3½ ells sinking and they had reached the Ryder Coal, to be
rewarded with 4*d*. (probably in ale). A further 3 ells provided them
with a further reward because of the toughness of the work and the
presence of water, and another 2 ells of this tougher work brought them
to the Seven Foot Coal and again 4*d*. reward. The depth of the shaft was
then 84½ feet, but they continued it until they were 1 ell below the coal,
making a recorded depth of 23½ ells or 88 feet.

More men were drafted into the partnership to make several headings
from the bottom of the shaft into the coal, but labour problems soon
arose and the employees were described as 'refusing and gevening up
their Bargayne' (App. p. 148). From then on Walker was the headman
and it was subsequently described as 'Walkers Pitt'. The favoured
position of the Griff miners is evidenced from this occurrence, as there
were no apparent repercussions to their broken contract. In north-east
England colliers were employed under much severer conditions and a
fine of £18 was extracted for such breaches of contract.[32] Other employees
were quickly found for Walker's pit, and heading continued, this time
with written contracts: 'in earnest to Bugs Walker Gaskin & Hawkins
for to dryve the heads as their bargayne is in wrytinge'.

Several heads were driven through the coal from the foot of Walker's
shaft in order to open up the seam to start working (it is not clear if these
heads are shown on Fig. 3). Four of these were north–south heads
parallel with the strike of the coal but the fifth was a head eight yards
'down the hill', that is, down dip, terminating at the 'crosse head in the
Deepe'. This cross-head was below the level of the water pit and water
had to be scooped up (presumably with hand-bailers) to that level
'untill Coales of the Deep be gotten' (App. p. 151). One would have
expected the miners to have sunk the water pit a few feet deeper to
drain the lowest head, rather than require it to be hand-bailed. The
whole sinking operation was hampered by water and several payments

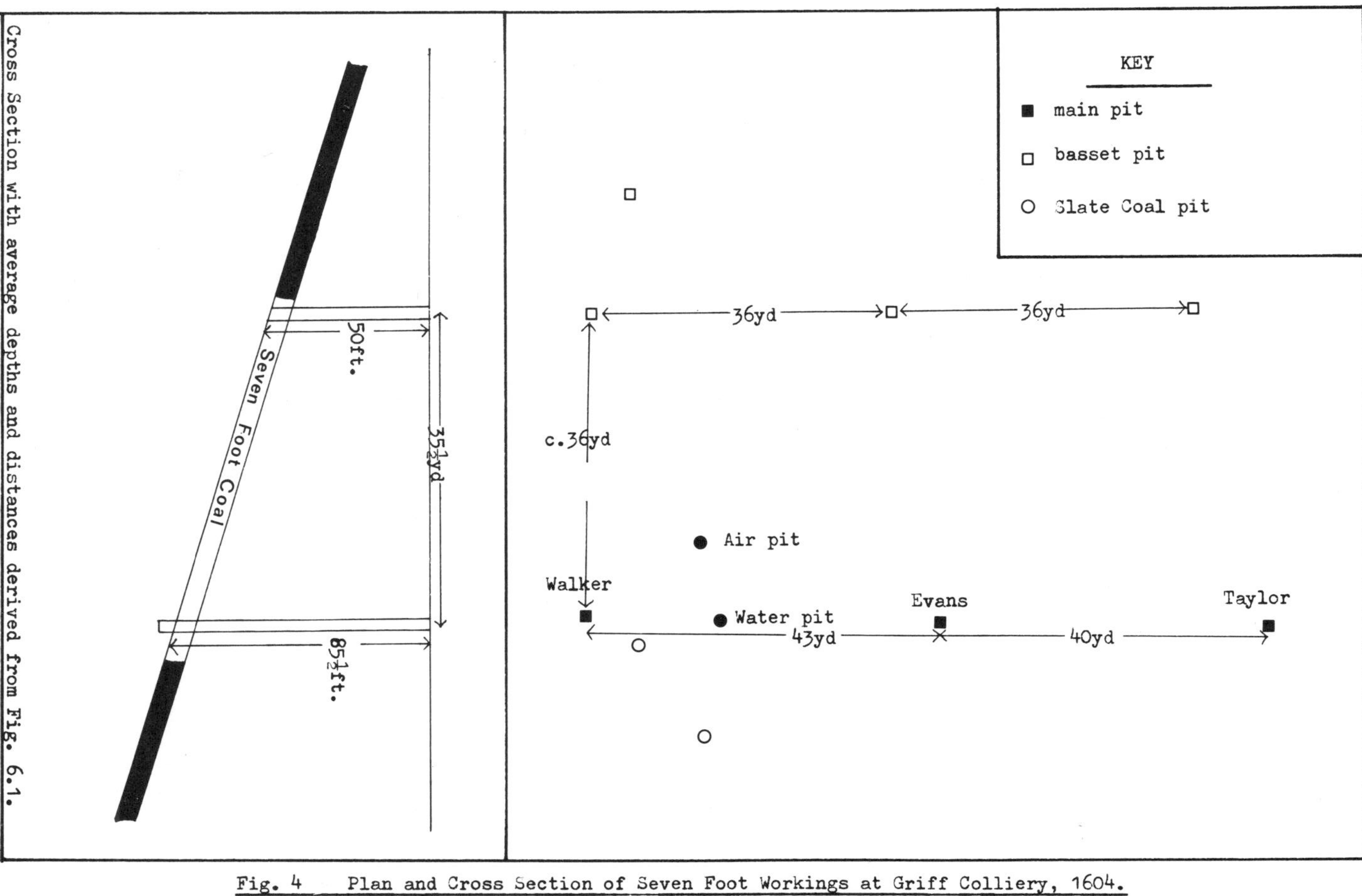

Fig. 4   Plan and Cross Section of Seven Foot Workings at Griff Colliery, 1604.

had to be paid to men to draw the water in the short lull between Taylor's demise and Walker's arrival (App. p. 149).

Labour was also required for 'making the turne barell [i.e. windlass] to turne the spoyle' in order to get waste out of the pit, while underground several 'paire of couplings' were laid. (Couplings or 'double timbers' consist of two upright posts and a crossbar supporting both the roof and sides of a heading.) These labours were successful and on 17 March 1604, 1*s.* was paid to 'Walker & his partners by Mr. Fyton for the sight of the First Coale'. Coal was not produced in saleable amounts, however, until 21 April, when the pit left the sinking phase and entered its production phase. The total costs of sinking and heading Walker's pit to 21 April were £26. 3*s.* 10*d.*

The other two pits that were being sunk simultaneously by Evans and by Butcher (later Taylor and then Kilby) display very similar patterns of sinking and costs. Evans's pit (86 feet deep plus a water sump under the coal) was producing coal by 28 April after initial costs of £22. 13*s.* 3*d.* and Butcher's pit (also 86 feet deep plus a sump) came into production during the week ending 5 May, after requiring £25. 8*s.* 2*d.* for sinking and heading.[33]

Sinking costs were not confined to making the three main pits, for during March and April four shallower pits were sunk averaging 50 feet in depth and £3. 16*s.* 2*d.* each to sink.[34] They were described as being sunk to the 'base coale' but no clue is given as to what constituted the 'base [or basse] coal'. Though several seams do lie below the Seven Foot, they are rather thin (except the lowest one, the Bench Coal) but there is no record of their being worked in the southern part of the coalfield. The most likely interpretation is that the 'base coal' meant in fact the *basset coal*,[35] that is, the coal as it approaches the outcrop, so that the four pits were basset pits, and this explains the shallowness compared with the main pits. It was normal to construct basset pits in conjunction with main pits (with an underground head connecting them) to give easier access to the coal faces as the mining moved up the seam. They also permitted better circulation of the air and provided a possible escape route.

The coal was worked up-dip in order to facilitate drainage, but nowhere in the accounts is the particular method of mining described. By the layout of the colliery the two main methods, pillar and stall and longwall retreating were possible. Pillar and stall was the main method used at that time, but the only mention of pillars is in the Slate Coal accounts, where a payment was made to Wootton 'for sinking a Bassett pitt to get old post of Coale'.[36] Longwall working was a Shropshire innovation, which though in use by this time, is not normally assumed to have reached Warwickshire until the early eighteenth century, when Sir Richard Newdegate introduced the method in his Griff colliery.[37]

4. *Drainage*

The many references to the construction of the water drainage system are symptomatic of the effort and expense required in opening up a medium sized colliery such as this. There were two main drainage constructions, the sough[38] and the water pit. Soughs normally have an exit to the open air at the lowest possible position, but the sough at Griff was connected to Taylor's old pit in the Slate Coal workings (Fig. 3) and water raised to the surface from there. The sough was only capable of draining the first 50 feet of the colliery's depth (via Taylor's old pit), but performed a very useful drainage function while the other pits were being sunk, and before the main water pit was constructed. Indeed, it was first called the 'water pit', but to avoid confusion with the main water pit it will be referred to here as the 'sough pit'. The sough frequently became blocked with sludge, and payment was made to 'Farkson for boring 2 bords [i.e. boards] with holes to ley [i.e. lie] in the sough to staie the sludge' (App. p. 132). Even after the main water pit was working, the sough pit was still used for drainage, probably by means of a hand windlass drawing buckets or barrels of water to the surface.

The water pit was commenced in the week ending 24 December 1603, when Evans and Yates received an earnest of 1s. to start on the pit and 4d. for marking it out (App. p. 137). On 1 February they were paid 3s. 'for bottominge their pitt' (App. p. 138), the recorded depth being 97 feet, with a sump at the foot for collecting the water. After a few initial problems with the timbering of the pit a 'hovel' was built over its mouth so 'that they may worke dry' (App. p. 139) and water was soon being drawn to the surface.

The drainage process was very labour intensive, requiring two shifts or 'turns' a day, with six, seven or even nine men on a shift. A typical weekly payment for this was 'Itm. for xiiii[to] [i.e. 14 turns] of vii[en] men iiii[s]i[d] the to[rne] . . . £2. 17s. 2d.' (App. p. 139). Additional men were also employed in drainage when the need arose, and on one occasion the banksmen (i.e. the men who stacked the coal on the bank on the surface) were pressed into service at the water pit (App. p. 139).

The only guide to the actual drainage equipment used is in a section of the accounts dealing with 'hopinge Barrels, Trunckes and other necessaries of the Smyth' (App. p. 177) the barrels and trunks being the components of two different drainage devices, presumably one at each pit. Beer barrels were frequently purchased at Coventry, and there are many entries for the costs of repairing these vessels. They were most likely used as simple water containers on the end of a rope wound by a windlass, or as a series of barrels fixed to an endless chain or rope similarly suspended from a windlass. This was exactly the same as a bucket gin driven by horses or a water wheel. The large number of barrels in use at Griff favours the idea that they were connected for continuous haulage, but this in turn demanded a considerable amount of power, usually supplied by horses. Horses are hardly mentioned in the accounts so perhaps the large number of men employed at the water pits were doing the work of horses.

Terminological confusion is easy in mining vocabulary as miners also used the word 'barrel' to mean the horizontal or turning piece of lifting machinery. For example 1*s*. 8*d*. was paid for 'mending a gudgin[39] for a torne [i.e. turn] barrell' (App. p. 177). Several turn barrels were made as well as tackling trees, which were horizontal bars for hanging heavy tackle from, such as chains, ropes and buckets.

The second drainage device utilized 'trunks' as its main component, and interspersed through the payments for repairing barrels are payments for 'hooping trunckes' (App. p. 177). These trunks were almost certainly tree trunks, split lengthways and hollowed out to make the vertical pipes of a rag and chain pump. An endless chain with rags or leather plugs — originally brasses — travelled up the pipes carrying water from the water pit sump in the process. Rag and chain pumps were commonly used for drainage in the seventeenth century and an inventory of machinery at nearby Hawkesbury Colliery in 1674 lists the constitutents of several rag and chain engines.[40]

Having reached the surface the water was led off in a gutter lined with clay to prevent the water seeping underground again. Surface water also presented a problem, and considerable effort was required for 'scowringe the ditch to turne water from the hollowes of the seaven foote delph' and 'castinge earthe to dame out water from the crope [i.e. outcrop] of the seven foote'.[41]

Ventilation was not a highly rated requirement in coalmines in the seventeenth century so that in many areas fire-damp and choke-damp were a constant danger. At Griff, however, an air pit was sunk simultaneously with the construction of the sough drainage system, these being the first two operations before the main shafts were sunk (see Fig. 3 and App. p. 131). The air pit was 50 feet deep and connected with the sough or level as it was called, and its early construction suggests that explosive gas (fire-damp) was a considered danger. On the whole the Warwickshire coalfield is free of gas, but the colliery under consideration was afflicted as there are two references in the accounts to miners being 'hindered' or 'troubled with the damp'.[42] The construction of the basset pits was beneficial as they encouraged a flow of air down the main pits and up the seam to the basset pits. The fact that two shafts per pit (or colliery subdivision) were usual in Warwickshire may explain the otherwise infrequent occurrences of gas. One problem that did afflict the coalfield and Griff in 1604 was the high risk of fires. In March of that year a collier was paid for 'fillinge up Wottons pitt to dampe the fire out'[43] (Wootton's pit was a basset surviving from the Slate Coal Workings) and a few weeks later the air pit had to be stopped with earth to extinguish a fire there (App. p. 132).

5. *Wood and Miscellaneous Charges*

Wood for timbering the pits was largely obtained locally on the estate; the accounts refer to 'caringe [i.e. carrying] timber from Cheeply Wood and poles from Printofts Close' (App. p. 188). These places were within half a mile of the pits and wood also came from Spring Kidden Wood, which still exists on the northern part of the estate, 1½ miles from the site of the colliery. Wood was also occasionally purchased from others, for example 'to Henry Holden for xxxiii poles for the colepittes, bought by Ro. Gee . . . 10s.' (App. p. 188). When timber was procured from the estate only the costs of cutting and transporting were entered — totalling £60 for the period — no purchase price or capital value being recorded. Though one recognizes that a special relationship existed between Sir John Newdegate and Leveson and Fitton, it was usual when leasing mines to allow free use of timber within the bounds of the leased area, providing a very valuable asset to the lessors, owing to the large quantities of wood required and its high cost in the seventeenth century. (Geoffrey Foxe did not have this concession at his Griff colliery and paid £150 for wood.) A variety of names is used for the cut wood, depending on its ultimate use, such as, poles, settings, couplings, principals and faggots (App. p. 188). Couplings were essentially posts and lintels for holding up the roofs of the galleries, while polings and settings were the everyday pit props, faggots were made from the branches and trimmings, and though used at the pits for roofing the pit-mouth shelters, they may also have been sold as firewood, though no sales are entered in the book.

Other uses for wood included making 'cofferinge timber' for holding back water, and 'axeltrees for Wyndes'.[44] An axle-tree is simply a wooden axle, but in coalmining terminology frequently means the vertical centre shaft of a horse gin.[45] This would suggest that horse gins (which are essentially horse-turned windlasses) were in use for raising the coal, as a skip (i.e. basket) of coal is much heavier to lift than water. One possible detraction from this inferred use of horses is the almost total lack of references to horses in the accounts, other than a small payment to a collier for burying one. It is possible that the feeding of the horses was at the expense of the estate, and so was not here recorded. If the headmen were responsible for the horses out of their own pockets it would again not be entered in the accounts.

The coal was raised in skips, costing 8*d*. each to make (App. p. 180) and securely nailed together, unlike the woven baskets or corves used in other areas. Ropes were expensive items; a visit to 'Watson the Roper for 2 Ropes . . . 29 yards appeece' cost £1. 6*s*. (App. p. 180) and this was repeated for each pit. A great variety of tools and other requirements were mentioned, either as purchases or repairs at the smithy. Shovels cost 1*s*. each new, 'iii gawne payles' (gallon pails) 8*d*. each, and 'an hundred of sixpeny nayles & other nayles' 1*s*. 6*d*. The 'repayringe of a matock' cost 8*d*. and of a shovel 2*d*. Picks, hoops, cleats, locks, wedges, gudgeons and brasses also figure in the inventory of hardware charges.[46]

By 21 April 1604 the colliery had started saleable production. No attempt was made in the accounts to distinguish between the capital costs up to this date, and the operating costs later on. However, it is possible to separate them by inspection of the entries, and this has been done in Table 1, where the first column of costs represents the capital investment during the sinking phase[47] and the second column the operating costs after 21 April when the first coal was sold. The total capital required was £181. 17*s*. 10*d*., part of this coming from the profits of the Slate Coal sales, the rest being obtained by William Henshaw 'Clark of the Colle Woorkes at Griffe' as imprest money from William Whytehall, financial controller of the colliery. The profits from the Slate Coal pits and the cash advances from William Whytehall resulted in a healthy balance of cash in hand and this may have created a false sense of security and allowed easy payments in operating the Seven Foot pits, which were in fact incurring charges far in excess of those recorded for the Slate Coal pits.

6. *Production*

A large section of the account book deals with production figures and subsequent sales on a week by week basis. Each team of miners had its own bank for stacking their coal, which was sold direct from the bank, so the production and sales figures were kept close together. For each week of the six months from 21 April to 27 October (28 weeks in all) details are given for the amount of coal raised by each team, how much was sold from their bank and how much remained on their bank at the end of the week. Appendix p. 89, for example, gives the production by Walker's team in the week ending 8 September, when they raised 97 three-quarters of coal, which with the 42 three-quarters and a half load (i.e. 'di^{lo}') already lying on their bank made a total of 129 three-quarters and a half load. One hundred and twenty three-quarters were sold from their stack during the week, leaving 9 three-quarters and a half load at the end of the week's activities. The charter payment to Walker and his partners for raising this coal is shown in Appendix p. 116. Ninety-seven three-quarters at 10d. a three-quarter earned them £4. 0s. 10d., plus 4d. extra for getting 20 three-quarters in one day (this was entered under the development charges of Walker's pit, App. p. 154).

The total amount of coal raised by all 3 pits during the six months was 3,670 three-quarters. In addition there were 79 three-quarters of soft coal, and though the men were paid the same wages for raising soft coal, it was not added to the banks as soft coal was virtually unsaleable then, usually being left mouldering in heaps, or carried away by anyone who wanted some. The average quantity of hard coal raised was 131 three-quarters per week, but the actual amounts fluctuated quite widely, as seen in Fig. 5. These variations in total production are the result of the fluctuating outputs of the three groups of partners whose individual outputs are illustrated in Fig. 6. The temporal pattern of output is the same in each of these three cases, that is, the peaks and troughs occur during the same weeks, with the result that these individual patterns are mirrored in the composite production pattern (Fig. 5).

In searching for an explanation of this varying production, uneven demand is a possibility, but as will be shortly demonstrated, sales of coal closely followed production rather than vice versa. The fluctuating output was largely due to production problems and other demands on the miners' labour. Judicious analysis of the accounts for the relative dates provides the evidence required. During the weeks ending 9 and 16 June, for example, production fell to one-third of what it had reached after a promising start, and is represented as the first trough in Fig. 6. Fig. 5 shows that Walker's pit produced no coal during these two weeks, and Evans's and Taylor's pits were also on a much reduced level. Far from enjoying an early summer holiday, the men were battling

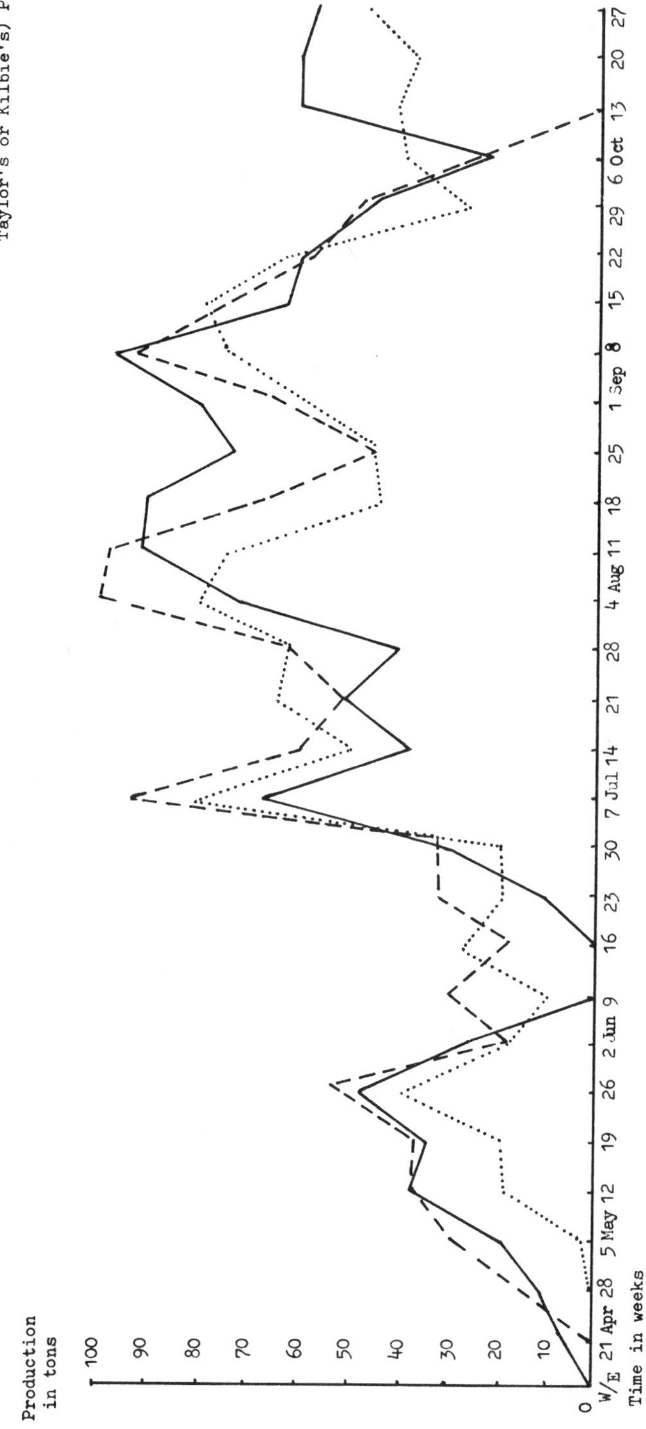

Fig. 5   Weekly Production from each pit at Griff Colliery from Week Ending 21st April to 27th Oct. 1604.

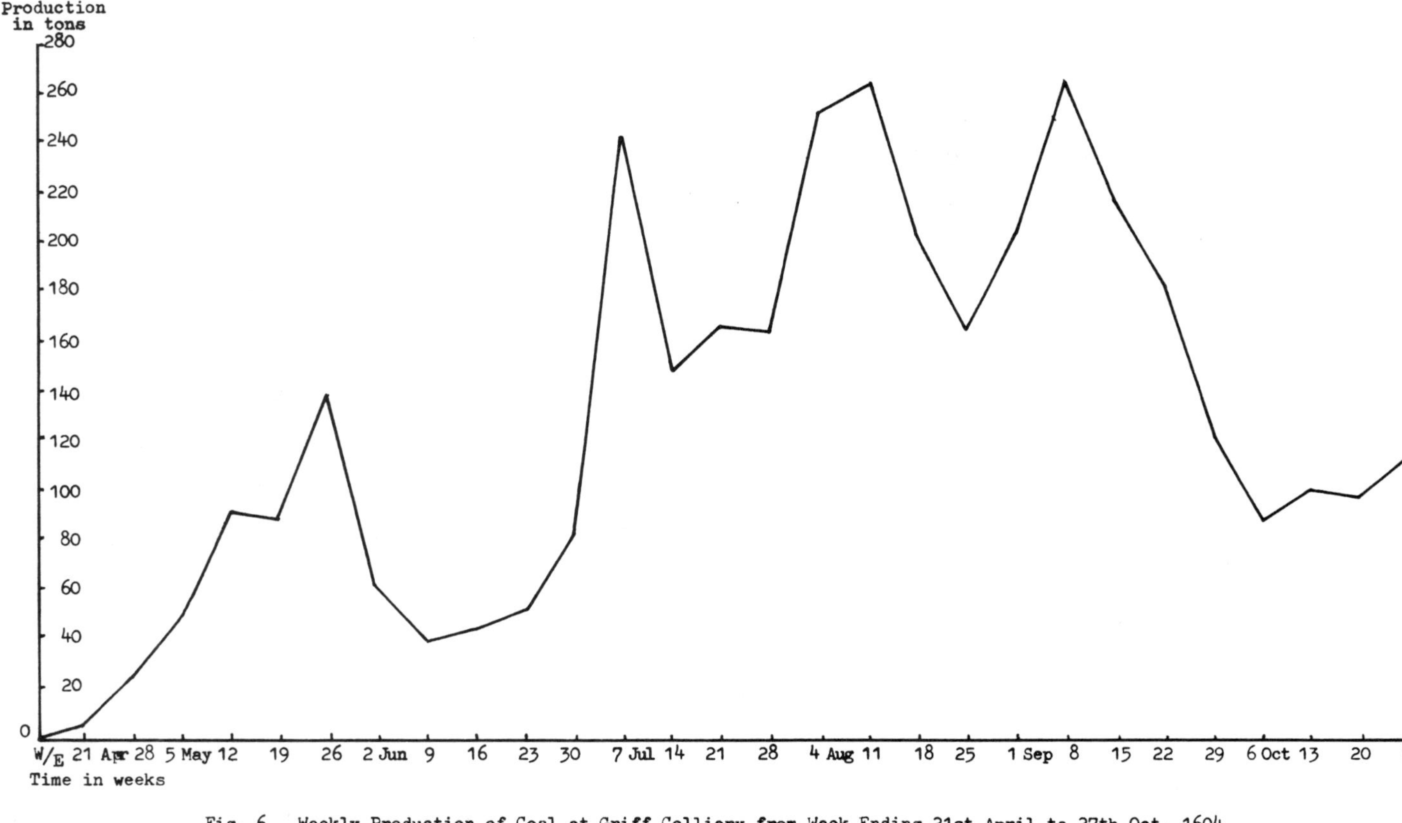

Fig. 6   Weekly Production of Coal at Griff Colliery from Week Ending 21st April to 27th Oct. 1604.

with an inrush of water in Walker's pit, as included among the charges for these weeks are payments for 'drawinge water by reason of the floud' and 3s. 4d. for 'timbringe and recovering his shaft' (App. p. 152). The proximity to Sir Thomas Beaumont's workings is evidenced in that section, and the floodwater appears to have come from his works as Walker was paid for 'recoveringe of his worke towards Sr Tho.' and a small payment was made to Sir Thomas's Watermen for their efforts in keeping the water down. Walker was partly compensated to the tune of £1. 15s. for his lost charter money of 10d. the three-quarters (App. pp. 152–53) and Evans and Taylor also received payments as their production has also been affected by the height of the water.[48]

The other reasons for reduced production have less dramatic backgrounds, but still underline the problems attendant on coal-mining in the seventeenth century. Production rose quickly again after the flood in early June, but in mid July work fell back and various entries show that gateways were being moved, the level of the bottom raised and new headings cut, suggesting that it was easier to start again above the water rather than pump it all out. Another trough in production occurred in the second half of August, and though Evans's and Walker's pits provide no evidence, Kilbie and his partners (who had taken over Taylor's pit) received 1s. 'in consideracion beinge hindered wth ye dampe'.[49]

The amount of coal being mined fell off quite dramatically during September, but this was largely due to the incipient exhaustion of the pits. Tests were made during that month 'for tryall to see how much the coale dipneth',[50] suggesting that the coal was giving out faster than expected. The relationship between the depth of pit and workable distance up dip is $d = \frac{h}{\sin a}$ where d = distance workable, h = depth of pit, and a = angle of inclination. Thus the smaller the value of sin a, the greater the workable seam distance, and the steep inclination of the seam in the exposed Warwickshire coalfield meant that compared with other less steeply dipping coalfields, the same shaft depth produced far less coal. The Seven Foot Coal pits were about 100 feet deep, and if, as appears to be the case, the dip was 1 in 3 (i.e. $a = 18\frac{1}{2}°$), then the workable distance from the shaft bottom was 316 feet. However, if the dip had been 1 in 12 (i.e. $a = 4\frac{3}{4}°$) then the distance up dip would be 1,204 feet, that is almost four times as much coal could be obtained from the same depth of shaft. The rapid exhaustion was responsible for Evans's pit being described in the week ending 13 October as 'factum est' (i.e. finished),[51] and shallow pits were then sunk at both Taylor's and Walker's pits (App. p. 154) (in addition to the basset pits constructed earlier) to help in the extraction of the coal nearest the surface.[52]

7. *Sales, Profit and Loss, and Closure*

A graph of weekly sales is shown in Fig. 7, and though this shows even greater fluctuations than the graph of production, the peaks and troughs occur in exactly the same weeks. Sales were closely dependent on production, and the correlation coefficient for the two sets of data is 0.80, which statistically demonstrates the strong positive relationship. Regression lines for the two variables, sales and production, are drawn in Fig. 8 and observation of their gradients again shows the strong correlation. (If the two sets of data had been totally unrelated, the lines would have crossed at right angles.) At several times demand was equal to and probably greater than supply. Fig. 9 shows the stocks of coal on the banks at the end of each week, and comparison with sales in Fig. 7 shows in this case an inverse relationship, that is, the peaks and troughs are not coincident. This is hardly surprising, as one expects high sales to be reflected in low stocks and vice versa, but on several occasions the banks were cleared out, suggesting that more coal could have been sold if it had been available. The pattern of production and sales during the two busy weeks ending 8 and 15 September 1604 is given in Appendix p. 89. The miners raised 264 three-quarters of coal during the first of these two weeks, the highest amount recorded (though the same figure was also reached on 11 August), which with stocks of 148¼ three-quarters from the previous week, gave a total of 412¼ three-quarters for sale. Sales were 374 during the week, with 37 three-quarters left on the bank. Evans's bank was totally sold off during the week, and probably also Walker's and Kilbie's, but as they were constantly adding coal, it was not uncommon to find some later mined coal lying on the banks at the end of the week.

The actual quantities sold are also interesting. Appendix p. 89 gives recorded sales of 10, 20, 40, 60 and 100 three-quarters, so the fuel was being sold in large rounded quantities, presumably to coal merchants, again suggesting considerable demand. Frequently the amounts were converted to loads (i.e. amounts in three-quarters $\div$ ⁴/₃) as it was marginally cheaper to buy by the load rather than the three-quarters. Prices, however, did not remain constant during the six months. The commencing price was 2*s*. 7*d*. a three-quarters, though this dropped to 2*s*. 4*d*. in early May, probably to attract interest in the newly opened colliery. Once this interest was created the price returned to 2*s*. 7*d*. a three-quarters and 3*s*. 5*d*. a load. In the week ending 30 June, when

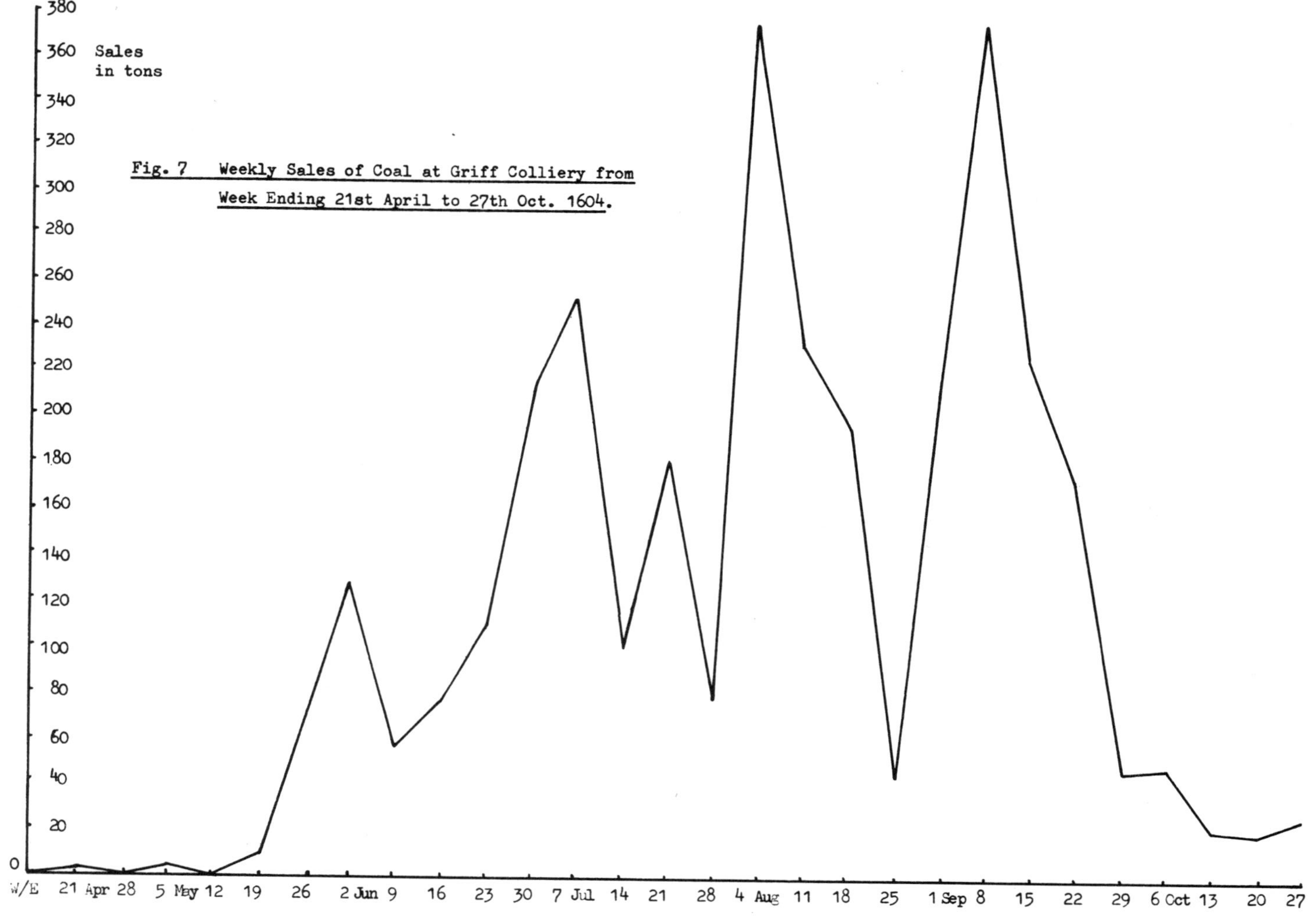

Fig. 7 Weekly Sales of Coal at Griff Colliery from Week Ending 21st April to 27th Oct. 1604.

sales were high, the price was suddenly raised by 19%, making the new tariffs 3*s*. 1*d*. a three-quarters and 4*s*. 1*d*. a load (App. p. 84). This had no effect on sales, and they were even higher the following week, again proving that demand was in excess of supply.

The selling prices were in keeping with others in the area. In the evidence given by Sir Thomas Beaumont in the Foxe case, he admits to having raised his price from 3*s*. 4*d*. to 4*s*. a rook (i.e. load) claiming that that was the price the other owners charged, and had been the price for

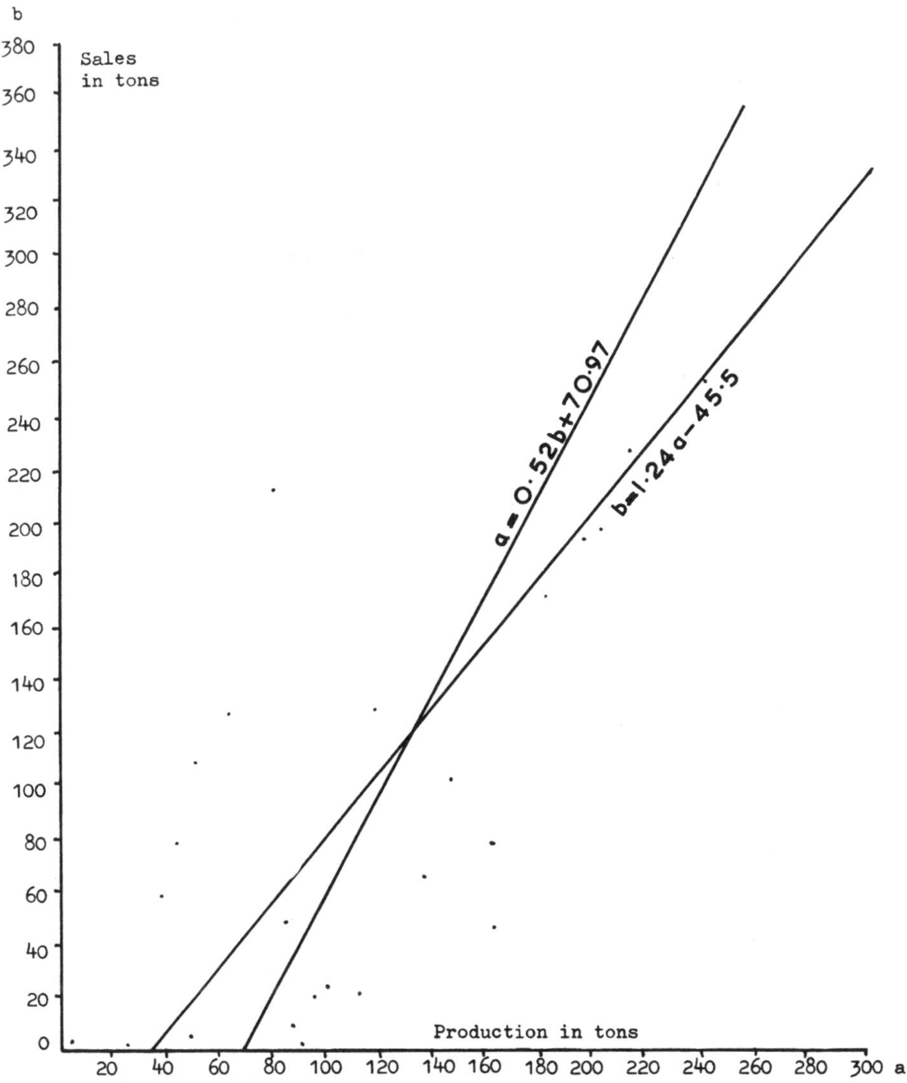

Fig. 8   Regressions of Sales and Production at Griff Colliery, 1604.

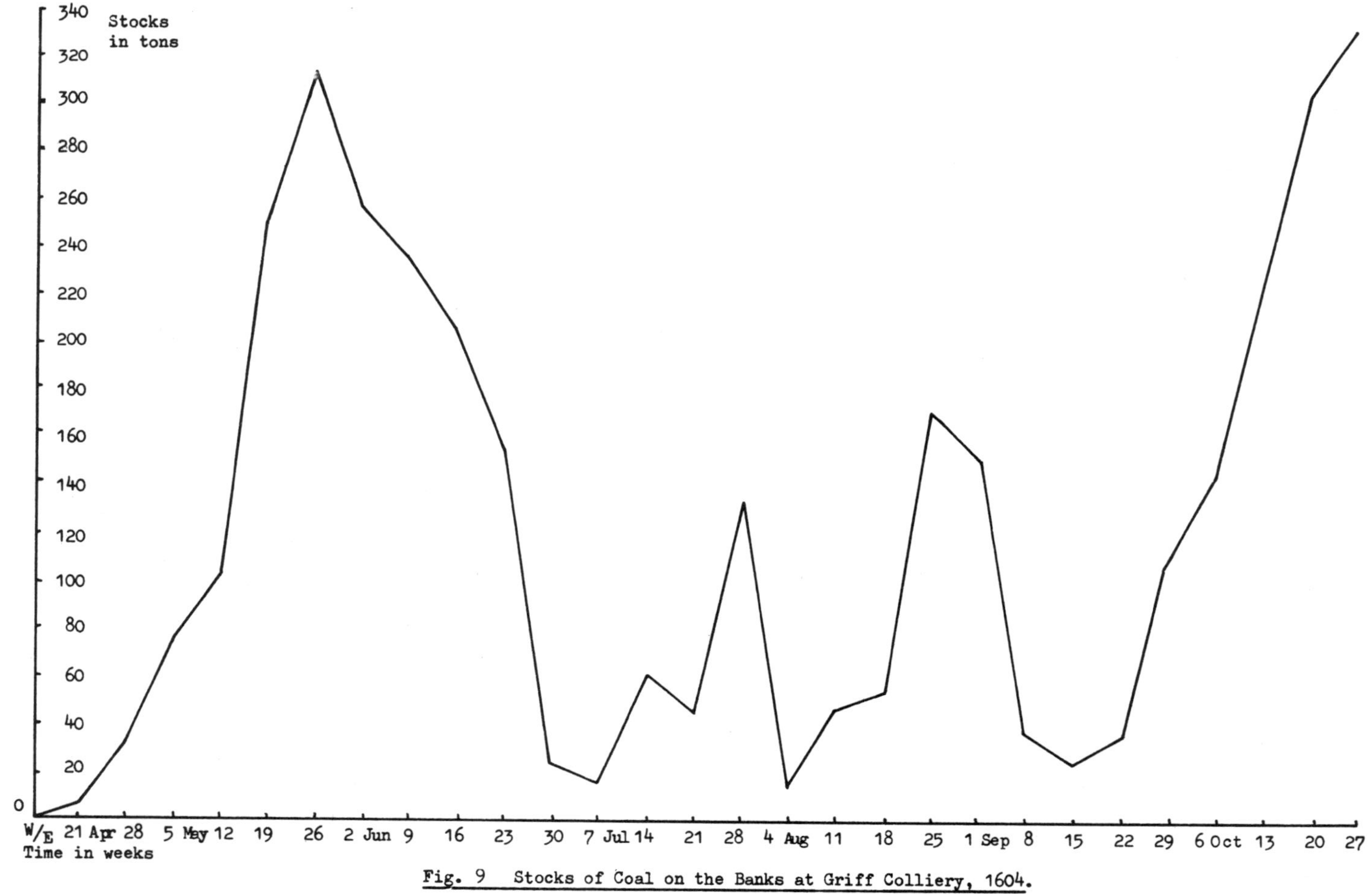

Fig. 9   Stocks of Coal on the Banks at Griff Colliery, 1604.

twenty years.[53] However, in the accounts of the working of the Slate Coal at Griff before the Seven Foot pits were started in 1604, the coal was sold for 2*s*. 4*d*. a three-quarters and 3*s*. 1*d*. a load.[54]

The sale of coal over the six months totalled 3,262 three-quarters and brought in an income of £483. 13*s*. 3*d*. Eighty-five three-quarters of coal were disposed of without any charge, mainly to Arbury Hall, under the terms of the lease (see App. p. 89 for quantities sent to Arbury) with small amounts to 'Burdes wief' and 'to Burne the Lyme Kill'.[55] The coal was transported from the colliery in carts which usually entailed frequent maintenance of the roads, as heavy loads soon churned up the surface. A labourer named Allen was paid 'for iii days mending the Laneway for the Cole Carts vi*d*. a day' and a further entry records a payment of 1*s*. 4*d*. 'to ould Allen for 2 dayes mendinge of Gryffe ford'.[56] The ford on Griff Brook had to be crossed on the road from Griff to Nuneaton.

On 27 October 1604 when the books were balanced and the main account ends, 330½ three-quarters lay on the banks, bringing the total quantity of disposable coal to 3,677½ three-quarters. This is near enough to the production figures of 3,670 three-quarters to underline the accuracy of the accounts and the close watch kept on the stocks.[57] The large amount of coal on 27 October is surprising and results from a dramatic fall off in sales after 22 September. No reason is given for this slump, but it is almost certainly explained by the coming of wet weather making the roads impassable. Until the arrival of the turnpikes (largely in the eighteenth century) cart loads of coal could only travel in the middle of summer when the roads were dry. It was sometimes possible to pull carts when the ground was frozen, but if the demand for coal was great enough during the winter (and it usually was, as poor people were unable to lay in large stocks during the summer) it was transported on the muddy tracks by packhorse. In the winter of 1684, when packhorses were used to carry coal from Hawkesbury Colliery to Coventry, the price of coal in the city rose from 7*d*. a cwt. to 12*d*. and 14*d*. a cwt.[58] July, August and September were the three best months of the short period for selling coal at Griff, followed by the very rapid turndown after the third week in September. Production also fell off during the period, but this appears to have been the result of incipient underground exhaustion, as already mentioned.

The balancing of the books on 27 October was a complicated affair, largely concerned with accounting for all cash transactions. The cash dealings for both the Slate Coal pits and the Seven Foot pits were added together, showing that expenditure on the pits came to £645. 16*s*. 0*d*. and receipts for coal sold to £693. 4*s*. 11*d*. (App. pp. 222–23). This small profit must have provided the operator with an unpleasant sight when he appended his signature to the books, as the Slate Coal pits alone had made a cash profit of £83. 13*s*. 11*d*. thus inferring that the new under-

Table 1

Seven Foot Coal Pits 19th Nov. 1603 — 27th Oct. 1604

Expenditure	19 Nov. 1603– 21 Apr. 1604			21 Apr. 1604– 27 Oct. 1604			Total		
	£	s	d	£	s	d	£	s	d
1. Trial pits in Smith's Close	1.	5.	6	—			1.	5.	6
2. Sinking (to 21st Apr.) & operating Walker's pit	26.	3.	10	14.	5.	1	40.	8.	11
3. Sinking (to 28th Apr.) & operating Evans's pit	22.	13.	3	7.	7.	0	30.	0.	3
4. Sinking (to 5th May) & operating Butcher's pit	25.	6.	9	17.	4.	11	42.	11.	8
5. Sinking four 'base coal' pits	15.	4.	8				15.	4.	8
6. Constructing and operating the sough	12.	15.	8	11.	8.	0	24.	3.	8
7. Making the air pit	2.	4.	2				2.	4.	2
8. Sinking and operating the water pit	39.	6.	6	48.	6.	10	87.	13.	4
9. Re-opening an old shaft and misc. surface charges e.g. levelling ground & mending roads	1.	15.	5	1.	0.	3	2.	15.	8
10. Surface expenses in Printofts Close	0.	14.	9	0.	11.	3	1.	6.	0
11. Blacksmith's charges	6.	16.	10	0.	10.	2	7.	7.	0
12. Ropes, tools and barrels	11.	17.	9	2.	19.	3	14.	17.	0
13. Wood charges (mainly wages for cutting)	8.	3.	9	51.	8.	3	59.	12.	0
14. Wages to Evans as surveyor	0.	18.	0	2.	11.	0	3.	9.	0
15. Watching the colliery field				1.	18.	3	1.	18.	3
							334.	17.	1[1]
Addl. charges at end of accounts:									
16. Additional wood charges	2.	3.	0	2.	10.	0	4.	13.	0
17. Wages to Gee & Bayliff for selling & measuring coal	4.	4.	0	5.	12.	0	9.	16.	0
18. Mr. Henshaw's expenses		4.	0		5.	0		9.	0
	181.	17.	10	167.	17.	3	349.	15.	1
19. Charter payments (i.e. wages for raising coal)							157.	19.	11
Total exp: 19 Nov. 1603 — 27 Oct. 1604							507.	15.	0

INCOME

	£	s	d
Total receipts 21 Apr.–27 Oct. 1604	483.	13.	3
Value of 330½ threequarters on the bank at 27 Oct.	51.	0.	4

[1] The total is given as £334. 16*s*. 1*d*. in the account book, but there is an error in the addition.

taking of the Seven Foot pits was operating at a loss. This cash loss can be easily obtained by abstracting the figures for the Seven Foot pits from the overall computation (see Table 1) showing that the total expenses to 27 October were £507. 15*s*. 0*d*. and receipts from sales £483. 13*s*. 3*d*., a loss of £24. 1*s*. 9*d*. It must be stressed, however, that these were only cash balances, and no consideration was taken of the value of unsold stocks of coal amounting to 330½ three-quarters and worth £51. 0*s*. 4*d*. Furthermore, the expenditure total included the high initial capital costs of opening up the pits. The colliery was given another six months to run, but unfortunately the detailed accounts for the next period were kept in a separate book, which is not known to

TABLE 2

Seven Foot Coal Pits 27th Oct. 1604 — 27th April 1605

	£	s	d
EXPENDITURE			
1. Operating charges	15.	10.	2
2. Charter payments	14.	13.	8
	30.	3.	10
INCOME			
1. Sale of 444 threequarters at 3*s*. 1*d*.	68.	7.	11
Cash Profit	38.	4.	1
2. Value of 238½ threequarters of coal on bank at 3*s*. 1*d*.	36.	15.	10

TABLE 3

Seven Foot Coal Pits — Total Accounts

EXPENDITURE		
1. First period	507. 15._0	
2. Second period	30. 3. 10	537. 18. 10
INCOME		
1. First period	483. 13. 3	
2. Second period	68. 7. 11	
3. Coal on bank at 27th April 1605	36. 15. 10	588. 17. 0
Profit		50. 18. 2

Source: W.C.R.O., CR136/V.147. (Abstracted from the details of income and expenditure.)

have survived, though an abstract of the chief sums was entered at the end of the existing account book (App. pp. 224–25). This second six months, to 27 April 1605, was a period of slow production and sales, 352 three-quarters of coal being raised and added to the bank and 444 three-quarters sold.[59] Operating costs were kept low at £30. 3s. 10d., and a cash profit of £38. 4s. 1d., was made during the term (Table 2). The profit and loss account for the Seven Foot pits for the whole period from 19 November 1603 to 27 April 1605 (when it appears to have closed down), shows that the total expenditure was £537. 18s. 10d., and total sales £552. 1s. 2d. In addition there were 238½ three-quarters of coal on the banks, worth £36. 15s. 10d., which must have been sold the following summer. This would have given a net profit of £50. 18s. 2d. (Table 3).

A £50 profit on such a venture was a poor return over eighteen months and it is doubtful whether the return justified the effort in running the colliery, in dealing with the miners and coping with underground problems and crises.[60] No attempt was made to cost the capital equipment transferred from the Slate Coal pits, nor the planning and overall direction of the colliery. Sir Richard Leveson, Francis Fitton and William Whytehall would have the role of directors in a modern firm and be in receipt of high salaries. If any such managerial costs were calculated, no indication is given in the accounts; not even the wages of Mr Henshaw, the colliery clerk, were entered, as he was probably paid directly by William Whytehall. If these managerial costs had been accounted for and a realistic rent paid, the venture would almost certainly have been a failure.

Comparison with more successful collieries demonstrated just how meagre the result was. In 1579 Bedworth Colliery operated by Willoughby and Beaumont, raised 5,187 rooks of coal (c. 7,800 tons) which was sold for £865, against an expenditure of £526,[61] giving a cash profit of £339 (not deducting capital depreciation etc.). Bedworth Colliery was about the same size as Griff Colliery, as the 3,670 three-quarters raised in six months at Griff can be considered equivalent to an annual production capacity of about 7,000 tons. Sheffield Park Colliery, Yorkshire, made £240 profit from July 1579 to December 1582, a profit of almost £70 per annum from an average annual output of only 1,200–1,300 tons, a much smaller undertaking than Griff.[62]

The poor result and the problems they had encountered were presumably behind a special agreement the operators concluded with Sir Thomas Beaumont. Early in 1605 Sir Richard Leveson and Francis Fitton agreed and covenanted with Sir Thomas not to get any coal within the lands of the Newdegates for five years, in return for £10 a year from Sir Thomas who also agreed to supply Sir John Newdegate with 80 loads of coal per annum.[63] Sir Thomas's main part of the bargain, however, was likewise to cease mining activities in Coton (presumably at Griff) during the five years, confining his interests to his other colliery at Bedworth, whence the 80 loads of coal would have been supplied to the Newdegates. Thus, at a stroke two important mining undertakings agreed to close and one feels the reasons must have been fairly compelling to require such action. Sir Thomas's colliery may also have been suffering from falling returns, and was certainly experiencing drainage problems.[64] Instead of rivalling each other at such close proximity the two knights may have decided to give in to economic forces. Sir John had an assured supply of coal to keep Arbury Hall habitable and Sir Thomas had his colliery at Bedworth — almost certainly a more successful venture — to keep his talents occupied.

The question of scale was a constant problem in running a colliery in the seventeenth century, as a balance had to be obtained between the capital investment and the expected production and income. The three size categories of collieries that can be identified during this period had different inputs according to their scale of operation. A small colliery, less than 5,000 tons output per annum, was viable so long as no major capital expenditure was required. A medium size colliery (5,000–20,000 tons per annum) of which Griff 1603–5 was an example,[65] needed an initial capital input of several hundred pounds (only part of which is discernible from the accounts) but the output in this case was insufficient return for the heavy development and drainage costs. If the Newdegate and Beaumont collieries had been one unit — as they almost were, both on the surface and underground — the costs of drainage would have been easier and the whole unit (still of medium size) would have stood a much greater chance of success. The operation of a large colliery, that is, more than 20,000 tons per annum, increased the chances for failure, especially if capital inputs of many thousands of pounds were met with erratic production and major drainage problems but if success was achieved, as Bedworth certainly did in some years, the returns could be high.

The modest colliery that operated at Griff from 1603–1605 neatly illustrates several aspects of colliery development in the Warwickshire coalfield during the seventeenth century. First, the general layout and developmental pattern of short-life pits was repeated for most collieries, the only difference being in the scale of operation rather than the process. Second, the continual day-to-day problems with labour, drainage and transport are evident, and third, the closure of the colliery was the result of economic forces that constantly troubled the county's early industrialists.

NOTES

[1] E. Grant (1977), *The Spatial Development of the Warwickshire Coalfield.* Unpublished Ph.D. thesis, University of Birmingham.

[2] J. U. Nef (1932), *The Rise of the British Coal Industry*, 2 vols, 2nd Imp. London, Frank Cass, 1966.

[3] Public Record Office (P.R.O.), S.P. Eliz., Vol. CCLXVI, No. 119.

[4] Nef was bent on making the point that the sixteenth and seventeenth centuries were of momentous importance in the growth of the coal industry, but the output figures he calculated for inland coalfields were potential outputs that assumed all the collieries were in production at one time, a highly unlikely state of affairs in the seventeenth century.

[5] Nef, p. 221.

[6] Grant, p. 146.

[7] The term 'mine' originally meant an underground seam of coal, so that the lease of a 'mine' does not necessarily imply that a shaft or working unit existed.

[8] Grant, p. 64.

[9] Grant, p. 65.

[10] P.R.O., E310, 26/149, fol. 50 and C66/1359 M.5.

[11] A. W. A. White (1970), *Men and Mining in Warwickshire.* Coventry, Coventry Branch of the Historical Association.

[12] R. S. Smith (1957), 'Huntingdon Beaumont: Adventurer in Coalmines', *Renaissance and Modern Studies*, 1, pp. 115–53.

[13] Grant, p. 89.

[14] E. Ekwall (1960), in *The Oxford Dictionary of English Place-Names*, 4th ed., derives the name Griff from Old Norse *gryfia* meaning a hollow or pit. The place-name Griff is first recorded in the later twelfth century, so it is likely that coal has been mined at Griff since then. Cf. *The Place-Names of Warwickshire* (1936) (English Place-Name Society, XIII), p. 80.

[15] Geoffrey Foxe, a former employee of Sir Francis Willoughby, leased some drowned pits at Griff and elsewhere from the Gifford family who owned part of the area. Foxe spent £600 on draining the pits, plus another £150 for timber, and established a complex system of drainage machinery at the site. Unfortunately he was unable to make any profit, or even pay his rent. Foxe was imprisoned for these and other debts, and during his incarceration Sir Thomas and Huntingdon Beaumont leased the mines from the Giffords and seized all Foxe's machinery (P.R.O., C2 Jas.I F4/53). By 1603 when the Newdegates began mining at Griff, Sir Thomas Beaumont was apparently successfully operating one of Foxe's drained pits.

[16] L. Stone (1950), 'An Elizabethan Coalmine'. *Econ. Hist. Rev.*, 2nd series 3(1), pp. 97–106.

[17] Warwickshire County Record Office (W.C.R.O.), CR136/V.147. The Slate Coal and Seven Foot Coal delphs were mines or diggings on or near the outcrops of the Slate Coal and Seven Foot Coal. About 27 seams of coal exist in the Warwickshire coalfield but only a handful are commercially exploitable. The Slate Coal and Seven Foot Coal are two important seams towards the base of the Productive Coal measures.

[18] M. Robbins (1953), *A New Survey of England: Middlesex*, London, Collins. Arbury Priory is now called Arbury Hall. In 1675 Harefield was conveyed back to the Newdegates and remained in their hands until the twentieth century. Cf. *V.C.H. War.*, IV, p. 176.

[19] S. E. West (1968), 'Griff Manor House (Sudeley Castle) Warwickshire', *Journal of the Archaeological Association*, 3rd series, 31, pp. 76–101. A fortified manor house existed there in the thirteenth and fourteenth centuries, but only an empty moat — known as Sudeley Castle — remains.

[20] *V.C.H. War.*, IV, p. 176.

[21] For example, an estate account book of 1688–89 (W.C.R.O., CR136/V.130) has colliery expenses intermingled with agricultural expenses, building, kitchen and household accounts.

[22] W.C.R.O., CR136/C/488.

[23] This is on the back of the 1604 plan of the colliery.

[24] W.C.R.O., CR136/C/494.

[25] 'Colliery' is used here to signify a collection of pits operated together as one production unit.

[26] W.C.R.O., CR136/V.147, p. 191.

[27] W.C.R.O., CR136/V.147, p. 79.

[28] A 'three-quarters' (of a load) was the usual measure of coal used in the accounts. Nef (1932), suggests a load is about 1½ tons, so that a three-quarters is approximately 1 ton.

[29] Nef, Vol. I, p. 415.

[30] A head or heading is the usual term for a horizontal tunnel taking off from a shaft.

[31] A binding fee was a customary payment to colliers in most districts.

[32] T. Robertson, ed. (1970), *A Pitman's Notebook; The Diary of Edward Smith, Houghton Colliery Viewer, 1749–1751*, Newcastle.

[33] W.C.R.O., CR136/V.147, pp. 157–62 (Evans's pit) and pp. 163–70 (Butcher's pit).

[34] W.C.R.O., CR136/V.147, pp. 205, 209, 213 and 217–18.

[35] No etymological derivation for *basset* has been found, but it is likely that it is derived from the French *basse* = low.

[36] W.C.R.O., CR136/V.147, p. 48. The actual payment is on pp. 66–67.

[37] White (1970), p. 31.

[38] *Sough* is a Midland word for an underground drain, usually in association with mining. The sough at Griff was 51 yards long.

[39] Gudgeons are the metal bars projecting from a horizontal drum and turn in sockets. These sockets or seats were often made of brass, presumably to reduce friction. References to brasses, however, can also mean the plugs used in a rag and chain pump.

[40] City of Coventry Record Office, Trade and Industry, 7.

[41] W.C.R.O., CR136/V.147, p. 193.

[42] W.C.R.O., CR136/V.147, pp. 168 and 213.

[43] W.C.R.O., CR136/V.147, p. 174. Wootton's pit was also known as the 'fyre pitt' (Fig. 3).

[44] W.C.R.O., CR136/V.147, p. 187.

[45] W. S. Gresley (1883), *A Glossary of Terms used in Coalmining*, London.

[46] W.C.R.O., CR136/V.147, pp. 177–82, passim.

[47] A considerable amount of uncosted capital investment must have been carried over from the Slate Coal pits to the Seven Foot pits, such as for drainage equipment and the organizational charges of the labour force.

[48] W.C.R.O., CR136/V.147, pp. 159 and 166.

[49] W.C.R.O., CR136/V.147, p. 168.

[50] W.C.R.O., CR136/V.147, p. 194.

[51] W.C.R.O., CR136/V.147, p. 91.

[52] Similar shallow pits had also been utilized in the extraction of the last of the Slate Coal. (W.C.R.O., CR136/V.147, p. 66).

[53] P.R.O., C2 Jas.1. F4/53.

[54] W.C.R.O., CR136/V.147, pp. 4–25.

[55] W.C.R.O., CR136/V.147, p. 90 and p. 83.

[56] W.C.R.O., CR136/V.147, pp. 175–76.

[57] An old man was paid 2*s*. 6*d*. a week for 'watchinge the ffeilde' (W.C.R.O., CR136/V.147, pp. 183–84).

[58] P.R.O., E134, 36 Chas II, Michas. No. 43.

[59] Calculated from £14. 13*s*. 8*d*. paid in charter money, and £68. 7*s*. 11*d*. received for coal sold.

[60] *V.C.H. War.*, II, p. 221, is quite wrong in its calculation of £70 profit and 8% return per annum for the Griff Colliery, as it uses intermediary balancing figures from the end of the account to calculate these sums, balances which have no real bearing on the profitability of the colliery. The authors, however, do preface their calculation with 'so far as can be made out . . . .'.

[61] University of Nottingham, Middleton MSS, Mi.Ac. 125.

[62] Stone, op. cit.

[63] W.C.R.O., CR136/C.493. Curiously enough, though the document was dated 27 February 1605 the agreement was to run from 27 October of the previous year, that is the end of the first six months of the Seven Foot production. That was clearly impossible retrospectively and the colliery does not appear to have closed until 27 April 1605. It is also possible that the agreement was in hand on 27 October 1604, and the reduced production and sales of the next six months may well be a reflection of the progressive running down of the colliery.

[64] P.R.O. (State Papers Domestic), SP16, Vol. 204, No. 83. Sir Thomas proposed a joint drainage scheme but it was turned down by the Newdegates (W.C.R.O., CR136/C.494).

[65] About 3,700 tons were raised in six months, so a productive capacity of approximately 7,000 tons per annum can be inferred.

APPENDIX

Selected Pages of the Original Accounts

Note: The pages are numbered at top right or top left as in the original document, and are thus referred to in the text. The pagination sequence for this Occasional Paper has been placed top left or right. Owing to lack of space in the original Seven Foot Coal Accounts, charter charges and production and sales of coal were placed at the end of the Slate Coal Accounts, before the Seven Foot Coal accounts begin. Sections a & b of this appendix should thus logically be placed between sections g and h.

a.	Production and Sales of Coal	pp. 84 and 89
b.	Charter charges	p. 116
c.	The sough	pp. 131–32
d.	The water pit	pp. 137–39
e.	Sinking and operating charges of Taylor's (and later Walker's pit)	pp. 147–55
f.	Charges of blacksmith and for tools	pp. 177 and 180
g.	Charges for wood	p. 188
h.	Final totals	pp. 221–25

The accounts on pp. 84 and 89 (and 79–108) of the original are laid out in a standard way:

Remains and amount mined	both	sold	Received	date of week ending	
(a) Coal remaining on each individual stack at the end of the previous week (b) Coal obtained during the week	a + b	(c) amount of coal sold from each individual stack during the week	money obtained for c	name of pitt	amount of coal remaining at end of week i.e. a+b−c

The Remaynes of Coales the 23th of June 1604

The price for 31 ^{3}qters in this weeke is after iii^{s}i^d a threeqters

(84 see) (note)

In this weeke the price for Coles weare raysed aterq   vendicii   L s d   The Remaines of Coales the 30th of June 1604

			L	s	d		
Walker Rem 67 di^{lo}) gotton this w. 31)	98 di^{lo} price ray.	88 ^{3}qters for 66^{lo} x ^{3}qters for	11 1	5 10	6 10	Walker	di^{lo}
Evans Rem 51) gotton this weeke 32)	83	80 ^{3}q for 60^{lo} Mr Bugs 1^{lo} To Arbury 1 ^{3}q	10 0 0	5 3 0	0 4 0	Evans	di^{lo}
Taylor Rem 35) gotton this weeke 19)	54 price raysed	12 ^{3}q^{lo}ters for 9^{lo} 15^{lo} for 4^{s}1^d 1 ^{3}q more	1 3 0	10 1 3	9 3 1	Taylor	21 ^{3}qters
Receaved for Coles from the 23th of June to the 30th)			27	19	9	xxvii^{Li}xix^{s}ix^d	

Remaynes of Coales the xxxth of June 1604						The Remaynes of Coales the 7th of July 1604	
Walker his pitt di^{lo} gotton this weeke 68 ^{3}q)	68 di^{lo}	60 ^{3}q for 45^{lo} 8 ^{3}q di^{lo}	9 1	3 6	9 9	Walker	0
Evans his pitt di^{lo} gotton this weeke 94 di^{lo}	95 a q	80 ^{3}q for 60^{lo} 3 ^{3}qters more	12 0	5 9	0 3	Evans	12 ^{3}q a q^{ter}
Taylor his pitt 21^{ters} gotton this weeke 81	102	80 ^{3}qters for 60^{lo} 12 ^{3}q for 9^{lo} 8 ^{3}qters more	12 1 1	5 10 4	0 9 8	Taylor	2 ^{3}qters
Receaved for Coales from the 30th of June to the 7th of July 1604			38	5	2	xxxviij^{Li}v^{s}ij^d	

Note: The first half of this page (84) was commenced in the original using old prices, but was replaced by a loose half page with the new prices, which has been reproduced above.

Remaines of Coales the ffirst of September 1604	uterq	vendicii	L	s	d	Remaynes of Coales the 8th of September 1604 89
Walker Rem 32 di^{lo} gotton this weeke 97	129 di^{lo}	100 ^{3}q for 75^{lo} 20 ^{3}q for	15 3	6 1	3 8	Walker Rem 9 ^{3}q di^{lo}
Evans Rem 18 a q^{ter} gotton this weeke 93	111 a q^{ter}	100 ^{3}q for 75^{lo} 10 ^{3}q for Arbury 1^{lo}	15 1 0	6 10 0	3 10 0	Evans Rem 0
Kilbie Rem 97 a q gotton this weeke 74	171 a q^{ter}	100 $^{3}q^{ter}$ for 75^{lo} 40 ^{3}q for 4 ^{3}q more	15 6 0	6 3 12	3 4 4	Kilbie Rem 27 $^{3}q^{ter}$ a q^{ter}
Receaved for Coales from the ffirst of September to the Eight of the same			57	6	11	Rec Lvii^{Li} vi^{s} xi^{d}
Remaines of Coales the 8th of September 1604						Remaines of Coales the 15th of September 1604
Walker Rem 9 q^{ter} di gotton this weeke 63	72 di^{lo}	60 ^{3}q for 45^{lo} 3 $^{3}q^{ter}$ & a qter	9 0	3 10	9 3	Walker Rem 9 ^{3}q a q^{ter}
Evans Rem 0) gotton this weeke 73)	73 q	60 $^{3}q^{ter}$ for 45^{lo} 2 $^{3}q^{ter}$ at	9 0	3 6	9 2	Evans Rem 11 $^{3}q^{ter}$
Kilbie Rem 27 a q gotton this weeke 79	106 a q^{ter}	100 ^{3}q for 75^{lo} Arbury ^{3}q a q^{ter}	15 0	6 0	3 0	Kilbie Rem 03 $^{3}q^{ters}$
Receaved for Coales from the 8th of September to the 15th of the same 1604			34	10	2	Rec^{d} xxxiiij^{Li} x^{s} ii^{d}

Receipte

116   September 1604   disbursmts onely for chalter
                                        of Coales

		L	s	d	
1th	Imp^r to Walker & his ptners for gettinge of 80 3qters at x^d a three q^{ter}	3	6	8	
	It to Evans for gettinge of 64 3qters at x^d a three qter	2	13	4	
	It to Kilbie for gettinge of 58 3qters at x^d a three q^{ter}	2	8	4	viii^{Li}viii^{s}iiii^d
8th	It to Walker & his ptners for gettinge of 97 3qters at x^d a three qter	4	0	10	
	It to Evans & his ptners for gettinge of 93 3qters at x^d a 3quarters	3	17	6	
	It to Kilbie & his ptners for gettinge of 74 3quarters at x^d a 3qter	3	1	8	xi^{Li}
15th	Itm to Walker & his ptners for gettinge of 63 3qters at x^d a threeqter	2	12	6	
	It to Evans & his perteners for gettinge of 73 3qters at x^d a three qters	3	0	10	
	It to Kilbie & his ptners for gettinge of 79 3qters at x^d a 3qters	3	5	10	viii^{Li}xix^{s}ii^d
	Suma Totalis pagi	28	7	6	

Sough or Levell	The Charges of the Sough beginninge the 19th of November 1603 by Ro Evans & others	L	s	d	131
19[th]	Imp for dryvinge xx[tio] yards from Taylors ould pitt towards the Seven Foote Delphe for iii[s] the yard	3	0	0	iii[Li]
26[th]	Itm for sinkinge the ffirst pitt called the eard pitt on the levell and for dryvinge 9 yards downe the hill to finde a was to geve pasadge to the water	1	8	8	xxviii[s]viii[d]
	December 1603				
3[th]	Itm to Evans & Partners for xii yards more in the Levell for iii[s]iiii[d] the yard towarde the water pytte	2	0	0	ii[Li]
x[th]	It for vi yards more in the [same] Sough iii[s]iiii[d] the yard towarde the water pytt	1	0	0	xx[s]
xvii[th]	Itm for iii yards more for iii[s]iii[d] the yard	0	10	0	x[s]
xxiiij[th]	Itm for vi yards more for iii[s]iiii[d] the yard	1	0	0	xx[s]
xxxi[th]	Itm for vi yards more for iii[s]iiii[d] the yard	1	0	0	xx[s]
9-18-8	*March 1603*				
3	It sendinge downe Timber 1[lo] to Evans mendinge the sough	0	0	4	iiii[d]
x[th]	It Hinnie iii[to] ix[d] a to[r] in cleansinge of the sough	0	2	3	ii[s]iii[d]
page	(Suma Totalis of the sough) (to the xxiii[th] of March is) pagi x[L]xv[d]　　　　ex	10	1	3	ex
exa.					

132

The Sough Chardges March 1604

		£	s	d	
31th	It to Rochell for dim a ^{to} feinge in the sough	0	0	3	iii^{d}
	Aprill 1604				
viith	It to Butcher Di^{to} Layinge tackling trees to put the water Downe the sough	0	0	6	vi^{d}
xiiijth	It to Burges for stoppinge the eard pitt with earthe to Dampe the fire out	0	0	4	
	It to Evans Burges Yates & Moore 11to apeece feinge the sough by wages xii^{d} a man the to	0	8	0	
	It ii boyes wth them 2to apeece vi^{d} a to a boy	0	2	0	
	It 2 Bankesmen over them turninge sluge iiii^{to} viii^{d} a man the to.	0	5	4	
out	It to Evans & others for feinge the sough through by bargayne of greate pmysed	1	15	0	
	It Jackson for boringe 2 bords with holes to ley in the sough to staie the sludge	0	0	2	
	It geven them in Ale deservinge it by following their woorke	0	0	4	
	It to Walker for 1to helpinge them when they wrought by wages	0	0	8	Li^{s}x^{d}
xxith	It Smythe for 1to bottominge in the sough 1to	0	0	6	
	It geven to the sough ffeyers whyle the sough was in ffeyinge by my M^{r} & my unckle in Ale	0	1	1	
	It Penifather for woorke in the sough	0	0	3	xxii^{d}
	May 1604				
6th	It to Evans for 1to makinge a dame for to stopp the water out of the Cole head in the lard pitt	0	1	0	xii^{d}
xiith	It to Evans for iij tor^{n} in Taylor ould pitt feinge & layinge tacklinge trees in the water pitt	0	3	0	
	It Yates 1 to dim doinge the same busines	0	1	6	
	It ffrances Moore in the licke 1to	0	1	0	
	It Christopher Wooton hewinge wood 11to for them	0	1	4	
drawinge water & makinge the gutter	Itm Hinnie Drawinge water their 1 tor$^{}$ ix^{d}	0	0	9	
	It Jackson 1to dim xii^{d} his boy 1to vi^{d}	0	1	6	
	It old Stockes 1to drawinge water vi^{d} making gutter iid	0	0	8	
	It young Stockes 1to drawinge water	0	0	6	
	It uncoveringe the pitt topp & coveringe it agayne to Baker	0	0	4	x^{s}vii^{d}
	Suma Tolis pagi	3	6	0	ex.

1603 December vith   137

<table>
<tr><td></td><td colspan="2">The whole charges of sinkinge of the)</td><td colspan="3"></td><td></td></tr>
<tr><td>Water Pitt</td><td colspan="2">water pitt beginne the 17th of December)</td><td></td><td></td><td></td><td></td></tr>
<tr><td></td><td>1603)</td><td>L</td><td>s</td><td>d</td><td></td></tr>
</table>

date	item	L	s	d	total
xxiiiith	Imp^r to Evans and Yates in earnest to sinke the water pitt	0	1	0	
	Itm for markinge their pitt	0	0	4	
	It to him for sinkinge viii Elnes for iii^svi^d ye Elne	1	16	0	xxxvii^siiii^d
xxxith	Itm to him for v Elne iiii^svi^d the Elne	1	2	6	xxii^svi^d
	Januarie 1603				
viith	It for drawinge the water whyle the sough was stroken in to the pitt	0	1	0	xii^d
xiiiith	It to Butcher More & Gest for every of them a torne & a halfe appeace sinkinge in the stone xii^d the Torne a man iii ^{qters} of a yard in ye stone	0	4	6	
	It to Joth Yates for a turne cuttinge the stone levell wth sough	0	1	0	
	It to Ward for 1to & di drawinge over them on the Bancke	0	0	9	
	Itm to Evans to woorke the place for torne stakes & to strike the barrell into the levell	1	6	8	xxxii^sxi^d
xxith	Itm to Evans for iii Elnes and di for vi^d the Elne	1	1	0	xxi^s
xxviiith	Itm to him and Yates for sinkinge of iii Elnes for vi^s the Elne xviii^s	0	18	0	
	Itm for hittinge the Ryder Coale	0	0	4	xviii^siiii^d
ffeb 1603 4th	It to them more for ii dim Elnes vi^sviii^d the elne	0	16	8	
	It to them vz Evans & Yates in souferance) of there hard bargayne by my Mr Appoynted)	0	6	8	xxxii^siiii^d
	pagi (ex pagi (* (	7	16	5	ex.
				*	

*   A sum of £8.14.9 had been entered here and then erased.

138

	L	s	d	
xith				
Itm to Evans for i Elne sinkinge & for makinge a trough to cary water from Taylors pit to the sough	0	12	0	
It hittinge the 7 foote Coale	0	0	4	
It for viii to of iiii men ii s the to	0	16	0	
It caundles for the woorkmen in the pitt iii Li	0	0	9	
It Penyfather ii to strykinge water to the sough	0	0	6	
xviiith				
It to them for ii Elnes & dim xiii s iiii d the elne	1	13	4	xxix s vii d
It keepinge the head feyed from Taylors pitt to the sough	0	2	0	
It to Evans Yates & Burges by my Mr his appoynt mt for bottominge their pitt by Thorsday night	0	3	0	
It for 11 to of iiii men ii s the torne	0	4	0	
				xlii s iiii d
xxvth				
Itm for vii to of iiij men ii s the torne	0	14	0	
It for vi to more of v men ii s vi d the torne	0	15	0	
It for a hurdle for the water pitt by old Stokes	0	0	6	
It Yates Butcher & Walker for ii to apeace in the pitt with timber beinge run in/xii d repay ringe the to	0	6	0	
It halfe a to more to Yates in the same busines	0	0	6	
It a banckesman ii to cutting timber vi d the to	0	1	0	
It Carter & Yates for setting of 2 peire of cuplinges	0	1	0	
It xii Li of Candles for the water pitt	0	3	6	
Itm in re to old Wootton for his opinion in the water pitt beinge then at ffault/by ye appoyntm t	0	0	6	
It geven to the water men a gawne of Ale by Roger for drawinge the water downe to sumpe the next day	0	0	4	
				xlii s iiii d
March 1603				
iiith				
It to Evans for vi to Clensinge and timberinge the water pitt	0	6	0	
It Hinnie for vi to w th him in the same business	0	4	10	
It Butcher & Carter iii to apeece xvi d the to	0	4	0	
It Drawinge the water xiiii to of v men vii d a ma the torne	2	0	10	
It Cuttinge timber for Evans being in the pitt settinge	0	0	4	
It a Jigg & a hurdle for the water men	0	0	7	
It xii ll of Candles by Roger	0	3	6	
				iii Li i d
pagi ((	8	14	4	
				ex

5-8-11

Water Pitt 1603   139

		L	s	d	
xth	Itm for ix^{to} of v men ii^{s}xi^{d} the to	1	6	3	
	It for v^{to} of vii^{en} men iiii^{d}id the to	1	0	5	
	It in Ale at severall tymes by my M^{r} geven	0	0	7	
	It to Taylor & Rowley for one day apeece in making a hovell over the pitt vii^{d} a man	0	1	2	
	It Candles vi^{Li} by Roger Gee	0	1	9	
	It to Rogers for makinge of 200 of fagots to Cover the pitt, that they may woorke dry	0	2	0	
					Lii^{s}ii^{d}
xviith	Itm for xiiii^{to} of vii^{en} men iiii^{s}i^{d} the to	2	17	2	
	It to Hinne by Roger Gee for strikinge the barrell iiii dayes more then vii^{d} a day	0	0	6	
	It geven in Ale to the Watermen by M^{r} ffyton	0	0	4	
					lviii^{s}
xxiiiith	Itm for xiiii^{to} of vii men iiii^{s}i^{d} the torne	2	17	2	
	It to Walker & Buges & Hawkins for layinge of xiiii peyres of Cuplings iiii^{d} the peire	0	4	8	
sumpinge	Imp^{r} to Walker for iii^{to} iii^{s} Buges iii^{to} iii^{s}) Hawkins i^{to} & a halfe xviii^{d} Burges ii^{to} ii^{s})	0	0	0	
	Yates ii^{to} ii^{s} Daniell i^{to} xii^{d} Marshall ii^{to} Ladinge xiii^{s}	0	13	0	
Bankesmen & Strikers	Imp^{r} Bankesmen to draw both the water & the feinge of the pitt whyle they sumped ffirst Hawkins iii^{to} xxi^{d} ffreman iii^{to} xviii^{d} Gaskin ii^{to} xvi^{d} Gilbert i^{to} viii^{d} ffox ii^{to} xvi^{d} Milner iii^{to} xxi^{d} Penifather strikinge the barell iiii^{to} xvi^{d} Lowse at Walkers pit i^{to} strickinge vi^{d} xi^{s}ii^{d}	0	11	2	
	It Hawkin one to^{r} more drawinge on the banke over the sumps	0	0	7	
	It to Hinnie for Woorke done at severall tymes by Roger Gee	0	0	6	
					iiii^{Li}vii^{s}i^{d}

		L	s	d
pagi	(	9	17	3

		L	s	d
exu – huc usque endinge at thancon 1604	(Suma Totalis of the water (pitt huc usque (26^{Li}/8^{s}ex	26	8	0

Taylors pitt 147

disbursed for Taylors pitt the 24th of
December 1603 Mr Walkers pitt

<table>
<tr><th></th><th></th><th>L</th><th>s</th><th>d</th><th></th></tr>
<tr><td>xxiiii th</td><td>Imp in earnest to him and Walker to sinke their pitt</td><td>0</td><td>1</td><td>0</td><td rowspan="3">xvi s iiii d</td></tr>
<tr><td></td><td>It markinge out their pitt</td><td>0</td><td>0</td><td>4</td></tr>
<tr><td></td><td>It to them for v Elnes sinkinge for iiis the Elne</td><td>0</td><td>15</td><td>0</td></tr>
<tr><td>xxxi th</td><td>It to Taylor for sinkinge of vi Elne iiis the Elne
    Januarie 1603</td><td>0</td><td>18</td><td>0</td><td>xviii s</td></tr>
<tr><td>xxi th</td><td>Itm to Taylor and Walker for sinkinge of iii Elnes for vis thelne</td><td>0</td><td>18</td><td>0</td><td rowspan="2">Lv s vi d</td></tr>
<tr><td></td><td>It for the head from their pitt to the water pitt beinge vi yards and a halfe iiis the yard</td><td>0</td><td>19</td><td>6</td></tr>
<tr><td>28 th</td><td>Itm in earnest to bind Homes Gaskin & Rowley to the pitt wth Taylor</td><td>0</td><td>1</td><td>0</td><td rowspan="3">xxii s iiii d</td></tr>
<tr><td></td><td>It to Taylor and his Partners for sinkinge of iii Elnes dmy for vis thelne</td><td>1</td><td>1</td><td>0</td></tr>
<tr><td></td><td>It hittinge the Ryder coale</td><td>0</td><td>0</td><td>4</td></tr>
<tr><td>ffeb 1603 4th</td><td>It to them for iij Elne more vis thelne</td><td>0</td><td>18</td><td>0</td><td rowspan="5">xxvii s i d</td></tr>
<tr><td></td><td>It garlandinge the stone water into a Corner of the pitt</td><td>0</td><td>2</td><td>6</td></tr>
<tr><td></td><td>Itm for ix Elne under the Stonne viiid thelne</td><td>0</td><td>6</td><td>4</td></tr>
<tr><td></td><td>& for halfe an Elne iiijd in Bargaine pmsed</td><td>0</td><td>0</td><td>3</td></tr>
<tr><td></td><td>It strykinge the water in their head di tor
    by Peter Sutton</td><td></td><td></td><td></td></tr>
<tr><td>xi th</td><td>It to them for ii Elnes vis viiid thelne</td><td>0</td><td>13</td><td>4</td><td rowspan="3">xvii s viii d</td></tr>
<tr><td></td><td>It allowed them iiii men 2 tor apeece vid a man the torne to Draw their water</td><td>0</td><td>4</td><td>0</td></tr>
<tr><td></td><td>It hittinge the viien foote Coale</td><td>0</td><td>0</td><td>4</td></tr>
<tr><td></td><td></td><td>6</td><td>18</td><td>11</td><td></td></tr>
</table>

148 ffebruary 18

Taylors pitt 1603
alis Walkers pitt.

		L	s	d	
xviiith	Imp^r for i Elne sinkinge to Tayler and his partners for vi^s viii^d	0	6	8	
	It to them for iiii to of iiii men to turne water ii^s the to^r	0	8	0	
	It geven in earnest to Taylor, Homes, Rowley Buges & Hawkins for the dryvinge of the heads	0	1	0	xv^sviii^d
25th 0-18-8	It for / It for i^{to} of iii men ii^s the to^r	0	2	0	
	It for halfe a yard in the head to Walker they refusing & gevening up their Bargayne	0	1	0	iii^s

March 1603

		L	s	d	
3th	It for ix^{tor} of v men ii^sxi^d the to^r before Walker undertooke his Bargayne for heads dryvinge	1	6	3	
	It to him for iiii yards in the head betwixt his pitt & the water pitt v^s the yard	1	0	0	
	It in earnest to Bugs Walker Gaskin & Hawkins for to dryve the heads as their bargayne is in wrytinge It before they undertooke their bargayne to Walker & Tayler for iii^{to} apeece sumpinge & makinge the turne barell to turne the spoyle xi^d a man the to^r	0	5	6	
3-11-5	liii^six^d *				liii^six^d *
xth	It to Walker for iiii^y more in the head from their pitt to the water pitt v^s the yard	1	0	0	
	It to them for 8 paire of Cuplings vi^d the cuplinge	0	4	0	xxiiii^s
xviith	It for vii^{en} Yards in the head towards Sr Tho: Beaumounts feild Levell wth water pitt iii^s the yard	1	1	0	
	It to him for vi yards from his shaft on the bassett towards Evans pitt iii^s the yard	0	18	0	
	It to him & his ptners for sumpinge of their pitt to the Levell of the water pitt	0	10	0	
	It Layinge of iii peire of cuplings vi^d a peire	0	1	6	
	It to Walker & his ptners by M^r ffytton for the sight of the ffirst coale	0	1	0	li^svi^d
		7	6	11	

* should read li^six^d

March 1603   Taylors pitt alis   149
                            Walkers 1603

		L	s	d	
xvii^{th}	Imp^r these paid by my M^r his appoyntmt beinge water Drawers when Taylor his ptnerg gave up their bargayne ffirst Cockes $i^{to}vi^d$ Hawkin ii^{to} xii^d Gaskin ii^{to} xii^d Carter ii^{to} xii^d Steyne ii^{to} xii^d Rochell 1to vi^d Penifather i^{to} iiii^d // $v^s iiii^d$	0 0	0 5	0 4	$v^s iiii^d$
xxiiii^{th}	Itm to him for iii Yards in his head below the Levell of the water pitt & done the hill $iii^s iiii^d$ the yard	0	10	0	
	It for halfe a yard more after v^s the yard havinge Rec^d for 8 already in the head betwixt the water pitt & theirs	0	2	6	
	It to him for ii yards in his head towards Evans pitt above the water pitt iii^s y^e yard	0	6	0	
	It Layinge a peire of cuplings	0	0	4	
	It Stockes for makinge 2 hurdles for his pitt to draw steikes uppon out of the heads /iii dayes	0	1	6	$xx^s iiii^d$
	$xxv^s viii^d$	1	5	8	
exa	Suma Totalis of this pitt is	15	11	6	

150   The Charges of Walkers pitt March 1604

		£	s	d	
31[th]	It to Walker & his ptners in consideration to sett out their worke in the Coale from the levell wall	0	15	0	
	It to him for v yards downe the hill iii[s]iiii[d] the yard	0	16	8	
	It for 2 yards in the Bassett above the water pitt iii[s] the yard / vi[s]	0	6	0	
	It to him for 1 yard in his Crosse head in the Deepe iii[s]	0	3	0	
	It to him for makinge a lade hole to lade the water to the water pitt sumpe	0	3	0	
	It 2 laders viii[to] apeece vi[d] the to[r] ladinge the water to the water pitt	0	4	0	xlvii[s]viii[d]
vii[th]	Aprill 1604 Itm to Walker & his ptners for xxiii yards in his Crosse head after iii[s] the yard	3	9	0	
	It 2 laders of water xiiii[to] vi[d] a to[r]	0	7	0	
	It Candles for the laders	0	1	0	
	It for iii ould Skopes to lade water w[th]	0	1	0	
	It to Walker & his Partners their beredge money by Bargaine pmised	0	10	0	iiii[Li]viii[s]
xiiii[th]	It to Walker & his ptners for openninge the water head to the souge & for timbring the same	0	6	8	
	It to him & his ptn[ers] for i[tor] drawinge downe the water in their pitt	0	4	0	
	It for 2 Axelltrees for their hoockes	0	0	4	xi[s]
xxi[th]	It to Walker & his Partn[ers] for xvii turns in his Crosse head on both sydes his pitt after iii[s]	2	11	0	
	It to them for feinge the earthe that was fallen in by reason their worke was rooffed with water	0	7	0	
	It for xi[tor] of 2 laders vi[d] a to[r]	0	5	6	
	It Candles to lade by	0	1	2	
	It for 2 Skopes viii[d]	0	0	8	
	It to them for a pere of cuplings laying	0	0	4	iii[Li]v[s]viii[d]
		10	12	4	

Walkers pittes Charges 1604   151

Aprill

Date	Item	£	s	d	Total
28th	It to him for standinge to the chardge of Ladinge of water untill Coales of the Deep be gotten, and to feye the water pitt sumpe	1	0	0	
	It to him for makinge a sumpe to draw water in his pitt by Bargayne agreed uppō	0	6	8	
	It to him for dryvinge of viii yards & a halfe for ii^s the yard from the deepe wall leavinge a post of coale for the water pitt on that syde twards Evans	0	0	0	
	It Drawinge water i^{to} of v men the pits being Thirled viii^d a mā	0	17	0	
	It to him for layinge Cuplings iiii paire iiii^d a paire	0	3	4	
	It to them for ffrydays Ale	0	1	4	
		0	0	8	xlix^s
vth	May 1604 — Itm to Walker for iii^{to} of v men 8^d a mā the to^r Drawinge water	0	10	0	
	It to them for ffrydayes Ale	0	0	8	x^sviii^d
xiith	It layinge a peire of cuplinges in their shafte	0	0	4	
	It for ffrydayes Ale	0	0	8	xii^d
xixth	It to them for ffrydayes Ale	0	0	8	viii^d
26th	It to Walker for latchinge betwixt Sr Thomas Beaumont his ground and ours for watchinge his pitt when Sr Thomas his men Thirled into o^r woorkes	0	2	0	
	It ffrydayes Ale	0	0	8	ii^sviii^d
		3	4	0	

152

Ann^o dmi 1604 June   Walkers Pit

		L	s	d	
2th	Itm to Walker and his fellowes their ffrydayes Ale	0	0	8	viii^d
9th	It for ix^{to} of v men at this pitt drawinge water 8^d a man xxx^s	1	10	0	
	It to Hawkins & Trever for feinge & Timbringe the head betwixt the water & it	0	1	4	xxxi^{s}iiii^d
16th	It for iiii^{to} of vii^{en} men 8^d a man drawinge water by reason of the floud	0	18	8	
	It to Jackson a collier for iii^{to} more then ordinary wages ii^d a to^r	0	0	6	
	It for viii^{to} more of iiii men 8^d a man the to^r	1	1	4	
	It to Walker & iii men more wth him for Timbringe & recovering his shaft running by reason of the floud	0	3	4	
	It in Ale to the water drawers	0	0	4	xliiiii^{s}ii^d
23th	It to Walker in pte of paimt for gettinge up & recoveringe his woorke beinge run in	0	15	0	
	It ffrydayes Ale	0	0	8	xv^{s}viii^d
30th	It to Walker & his partners in parte for recoveringe of his worke towards Sr Tho.	0	10	0	
	It to them for ffrydayes Ale	0	0	8	x^{s}viii^d
		5	2	6	

July 1604   Walkers pitt Charges   153

		L	s	d	
7th	It to Walker his last paimt agreed uppon for recoveringe up his woorke	0	10	0	
	It to Starborow their bottomer by my M^r his appoyntment	0	0	3	
	It their ffrydayes Ale	0	0	8	x^{s}xi^d
14th	It to Walker & his ptners for cuttinge their gateway & raysinge their bottom	0	5	0	
	It ffrydayes Ale	0	0	8	v^{s}viii^d
21th	It to Walker for ffrydayes Ale	0	0	8	viii^d
28th	It to Walker for cuttinge up his gate way	0	5	0	
	It ffrydayes Ale	0	0	8	v^{s}viii^d
4th	August 1604	1	2	11	
	Itm to Walker in reward for his paynes taken heretofore in surveinge the woorkes	0	15	0	
	It to him & his ptners their ffrydayes Ale	0	0	8	
	It to them for gettinge xx^{tie} one day	0	0	4	
	It to Starborow their bottommer by my M^r geven	0	0	2	
	It to Sr Tho Beamonts watermen in re geven by my M^r	0	0	4	xvi^{s}vid
xith	It to Walker & his ptners for gettinge 40 ^{3}q in ii dayes	0	0	8	
	It to them their ffrydayes Ale	0	0	8	xvi^d
	Suma To^{lis}	2	1	9	

154

Auguste 1604 — Walkers pitt Charges

Date	Item	L	s	d	Subtotal
25th	It to Walker & his ptners for ffrydayes Ale	0	0	8	viii^d
1th	September 1604 It to Walker & his ptners for takinge up the timber of a bassett pitt betwixt their pitt & the lane heretofore suncke	0	5	0	
	It for cuttinge up their gateway	0	3	4	
	It for gettinge 20 3qr one day	0	0	4	
	It to them for ffrydayes Ale	0	0	8	ix^s iiii^d
8th	It to him & his ptners for gettinge 20 ^3qters in one day	0	0	4	
	It to Lawkins by my M^r geven for goinge to Polesworth for Colliers	0	0	6	
	It to them their ffrydayes Ale	0	0	8	
	It to them				xviii^d
15th	It to Walker & his ptners in earnest to sinke a bassett coale pitt for xlvi^s viii^d	0	1	0	
	It to them in parte of paymt of the afforesaid Summe for their bassett pitt	0	6	8	
	Itm ffrydayes Ale to them	0	0	8	viii^s iiii^d
22th	It to Walker & his ptners for a meeter wages a weeke	0	4	0	
	It to them in pte of paymt for their Bassett pitt sinkinge	1	0	0	
	Itm ffrydayes Ale	0	0	8	xxiiii^s viii^d
29th	It to Walker & his ptners in pte of paymt for sinkinge their Bassett Pitt	0	5	0	
	It to them for a meeter his wages a weeke	0	4	0	
	It ffrydayes Ale to them	0	0	8	ix^s viii^d
	October 1604				
6th	It to Walker and his ptners their last paymt for sinkinge their Bassett pitt	0	15	0	
	It to them for a meeter his wages a weeke	0	4	0	
	It for ffrydayes Ale to them	0	0	8	xix^s viii^d
	Suma To^{lis}	3	13	10	

2-4-6

October 1604

Walkers pitt Chardges

155

		L	s	d	
13th	Itm to him & his partners their ffrydayes Ale	0	0	8	viii d
20th	Itm to him & his partners for ffrydayes Ale	0	0	8	viii d
27th	Itm to him & his partners for layinge 3 peire of Cuplinges in the shaft	0	1	0	
	It them for ffrydayes Ale	0	0	8	xx d
		0	3	0	
	Summa To^{lis} of the Charges of Walkers pitte from the xxiiii th of December 1603 unto the 27 th of October 1604 excepte chalter payment for coales out of his pitte as may appeare	40	8	11	xl tie viii s xi d

	The Charges for hopinge Barrels Trunckes & for other necessaries of the smyth	L	s	d	177
January 1603					
21ᵗʰ	Impʳ for hooping of tow trunckes	0	3	4	
	It for leinge (?) a mandrill & mendinge a shovell	0	0	10	
	It for hoopinge tow Barrels	0	14	6	
	It hoopinge and bewlinge (?) tow trunckes	0	5	6	
	It for hoopinge tow barrels	0	10	6	
	It for platinge 3 peales*	0	0	8	
	It for a new pycke and steelinge 3	0	1	2	
1-16-6	March 1603				xxxviˢviᵈ
iiiᵒ	It for hoopinge a Barell wᵗʰ i new hoope & iii old hoopes	0	2	6	
	It hoopinge a Barell wᵗʰ foure ould hoopes	0	1	0	
	It hoopinge & nosinge iii peales to lade wᵗʰ	0	1	10	
	It for hoopinge of iii Barrels wᵗʰ old hoopes	0	3	4	
	It hoopinge of 2 new Barels wᵗʰ new hoopes	0	14	0	
	It for 3 pickes viiiᵈ a peece iiˢ 2 longe wedges iiˢ	0	4	0	
	It sendinge to the smythie for slypes by Stockes	0	0	2	
3-3-4					xxviˢxᵈ
xviiᵗʰ	It for mendinge a gudgin for a torne barrell	0	1	8	
	It for v peire of hoockes and Truncke peares	0	1	4	
	It for showinge (?) a shovell	0	0	6	
	It for a truncke bewle	0	1	4	
	It for Loopinge a barrell wᵗʰ new hoopes	0	7	0	
	It for iii peire of gudgins xxˢ iii peire of Brasses iiˢiiᵈ a peire	1	6	6	
	It for iii peire of hoopes	0	3	0	
					xliˢiiiiᵈ
	Summa Toˡⁱˢ vˡiiiiˢviiiᵈ	5	4	8	

* i.e. pails

180   Extraordinarie charges for tooles ropese & such licke

	March 1603	L	s	d	
xvii[th]	Itm sendinge to the roper by Norman for ropes	0	0	6	
	It geven by my M[r] in re to Nicho Cobler by my M[r]	0	0	6	
	It to Robinson of Coton for tow peires of skipes	0	2	8	
	It for Barrell nayles by Roger Gee	0	0	6	
	It to Rogers for tow hundred fagots	0	2	0	
	It Asbie for one hundred makinge	0	1	0	
	It for 2 ropes by Roge[r] Gee to Watson xiiii[s] a rope	1	8	0	xxxv[s]ii[d]
xxiii[th]	It for 8 barrels by Roger Gee xii[d] a (barrell)	0	8	0	
	It for Cariadge of 4 from Coventree	0	0	8	
	It repayringe of a shovell to Ridley of Eaton	0	0	2	
	It Suttons wiefe riddelinge iii dayes	0	0	8	
	It for a shovell for Norman & Bucke to worke w[th]	0	1	0	x[s]vi[d]
	Suma To[lis] for these necessaries is vii[li]ix[s]v[d]	2	5	8	
xxxi[th]	It for Barrell nayles to the Naylor	0	0	9	
	It carringe a Barrel to Atleborow by Lowse	0	0	2	
	It to Watson for iii ropes xviii[s] the rope	2	14	0	
	It to Wotton for goinge to the roper to hasten w[th] them	0	0	6	lv[s]v[d]
	Aprill 1604				
xiiij[th]	It for a hogshead to make a barrell	0	1	0	
	It for a shovell to Roger Gee	0	1	0	ii[s]
xxi[th]	It to Watson for 29 yards appeece in 2 Ropes	1	6	0	
	It nayles for Barrels by Roger Gee	0	0	5	
	It for iii ridles ii[d] appeace	0	0	6	xxvi[s]xi[d]
xxviii[th]	It to Rayment the Cowper for makinge of 2 Barrels of 2 hogsheads ii[s] a barrell	0	4	0	iiii[s]
	The hole summe of this syde is vi[li]xiiii[d]	6	14	0	

188   The Chardges of hewinge of wood
                        Cariadge fallinge & such licke

	Aprill 1604	L	s	d	
28th	It to Roger for iii dayes ii^sviii^d the day Caringe timber from Cheepley Wood & poles from Printofts Close	0	8	0	viii^d
	May 1604				
5th	It to Roger Gee for iii dayes ii^sviii^d a day Caringe the toppes of trees for wast wood & Furr fagots to Cover the pits	0	8	0	
	It Wootton Cuttinge levell wood one daye	0	0	10	viii^sx^d
xiith	It to Roger Gee for ii dayes Caringe wood) out of Shipley wood ii^sviii^d the day)	0	5	4	
	It to Jackson for vi settinges hewinge vi^d	0	3	0	
	It for Cuttinge level wood one day to old Wootton	0	0	10	
	It Poles for the Colepittes by Richard Clowes in the Stockingford bought	0	6	10	xvi^s
19th	It Jackson for hewinge one settinge vi^d	0	0	6	vi^d
1-13-4	**June 1604**				
16th	It to Jackson for hewinge pollings & Principalls for a pitt to be sunke to recover the sough	0	2	0	ii^s
23th	It for fallinge & lowinge of 33 pooles of one Houlden bought by Roger Gee	0	1	6	
	It to Henry Houlden for xxxiii poles for the cole pites, bought by Ro. Gee	0	10	0	xi^svi^d
	July 1604				
21th	It to Jackson for hewinge of 8 peire of cuplinges	0	1	4	
	It to ffoster for iii dayes fagottinge &) lowinge for Colepittes in Springe Kiddinge)	0	1	9	iii^si^d
28th	It to Jackson & Peeter for ii dayes appeece cleavinge wood for the pites	0	2	8	
	It to Willin Lucas for 62 poles	0	7	0	
	It a boy for walkinge Mr ffyttons horse	0	0	2	ix^sx^d
	Summa Totalis	2	19	9	

Receipts

	L	s	d
Receyved for Coales uppon the Slate Cole Delphe) from the vi th of August 1603 unto the vth of) November 1603 as may appeare in that booke)	209	11	8
Alsoe received for Coales uppon the Seven foote) Delphe gotton from the xxith of Aprill 1604) unto the 27th of October 1604 as the booke) for the same delph weekly sheweth)	483	13	3
Also received at severall tymes and dayes) of my unckell Willm Whytehall as soe much) by way of imprest for and about the said cole) woorkes as doth appeare) in the first leafe of the ffirst booke	332	0	0
Totall Receaved	1025	4	11

Disbursmts

whereof

	L	s	d
Whereof the said Willm Henshawe prayeth) to be allowed as soe much layd out and paid) for and about the said coleworkes, and by him) expendyd about the same workes wthin the tymes) above menconed as by this Booke & Tow) other bookes keepte there of may particulerlie) appeare beinge	591	9	0
Alsoe wages to Roger Gee and Bayliffe for the) recaipte of moneys for coales and soe to the) stackinge and measure of coales, from the) 30th of July 1603 untill the 27th of October 1604) beinge Three score and ffoure weekes at iiis the) weeke	12	16	0
Also for viien weekes dyett for wth Henshaw vz from the vith of August 1603 untill the 24th of September at iiis the weeke	1	1	0
Also for Wm Henshawes dyett for a month setting all tymes togeather he continewed at Gryffe all Day in imploymt for the cole woorkes	0	12	0
Also more for moneyes lent to the colliers) at severall tymes as may appeare in a booke) at lardge from the vith of August 1603) until the 27th of October 1604)	11	5	0

222

	L	s	d
Also more to^e Roger Gee for xxiii^tie load of Poles for the pits at v^s the Load	5	15	0
Also more to him ix loads of fagots at ii^s the Load	0	18	0
Also more to Robte Smyth of Gryffe) in recompence for appeece of ground w^ch) he lendeth by lease called the Pingle to) gett the coale on the Slate Delph their) beinge, agreed uppon by my M^r S^r Jo. Newde-) gate)	2	0	0

	li	s	d
Sume Total of all) manner of & charges and) dysburcementes expended) for and aboute all their colle) woorkes and conteynyd in) theys three severall bookes) whiche now are heare all fixed) together) untill theys xxvii^th day) of October 1604)	625	16	0

whereof

Whereof

	l	s	d
Delivered by me Willm Henshawe as soe) much ready money to the hands of my) unkell Willm Whytehall betwixte the) tymes vz vi^th of August 1603 and the) v^th of November 1603 the Summe)	209	11	8
And also more paid for the cariadge of woode to the Cole pits betwixt the tymes afforesaid	1 20	0	0

Sume Total of) viii^C lv^l vii^s viii^d
all theis dysbursments)

So the said accomptant Willm)
Henshawe upon theis his accomptes)
standeth in debet and owinge this) Clxix^l xvii^s iii^d
xxvii Day of October 1604)

wch said debet of 169^l
17^s/3^d standinge charged upon Willm
Whytehall in this place as may apeare is
since the wrytinge of this accompte accompted
and <u>cleared</u> by the said Willm Whytehall upon his
accompte taken the 4 daye of November 160ỹ (1604 or 1606?)
and subscribed by the hande of Sr Jo. Newdigate
and ffr ffyton

(This last paragraph has been scored through)

223

Whiche said debet of of CLxix^{l}xvii^{s}iii^{d} the
said accomptant Willm Henshawe hath payed
over and delyvered the xiiith of November
1604 to the handes of his unckle Willm Whythall
And the said Willm Whythall is nowe to be
charged there bye and is accomptabell for the
same, And so the said Willm Henshaw is
now dischargyd thereof and touchinge theis
accomptes in theis 3 bookes apearing standeth cleare
                    this xxviith daye of
October 1604 conserninge
the same three books   Syned   Jo. Newdegate
                            francys
                            fyton
And the said Willm Whytehall hath synce accompted for the
said debet as in thend of the paige nexte heare apearing
maye apeare. This examyned by me ffr ffyton this
xix of Januarie 1606 francys
                        fyton

Upon advysed perusall and due examynacion hade
of all the contents included(?) in this booke beinge
the booke of accompte of Willm Henshall Clark of the
Collewoorkes at Griffe aswell for all summes of money
by hym Received upon   Imepreste by thande of his
uncle Willm Whytehall, as also all other summes of
money come to his handes for any colles sould at the
same collewoorkes betwene the xxxth daye of July 1603
and the xxviith daye of October 1604   viz
            (The money wch this accomptant   )
Receiptes   (hathe Received within the said tymes) 332^{l})
            (by waye of Imepreste$_{t}$ is —————————)   ) Mxxv^{l}iiii^{s}xi^{d}
            (And all the moneys y^{t} he hathe) l s d   )
            (Received for any Colles soulde)693/4/11   )
            (is   )   )
            The said accomptant apeareth by   )
            this same booke to have paide   )645^{lj}/16^{s})
            & disbursed in an aboute the said)   )Mxxv^{l}iiii^{s}xi^{d}
            Collewoorkes S^{m} Redy money   )   )
            And in lyverey money he the said   )   )
            accomptant apearethe to have del.(?))) l s d)Et sic Eque
            to thande of his uncle Willm   )379/8/11)huc usque
            Whytehall who hathe since made   )   )
            also his accomptes for the same   ) francys
                                                   fyton

Et Eq

Mem I find by my   Grandfathers Almanack he was alive Anno 1609,
however his uncle Fitton came to take this account.
    (in a different and later hand).

224

And also then upon the lyke parusuall and examinacion of one
other later booke of accompte of the said accomptante Willm
Henshaw consernynge the said colleworkes in Gryffe
aforsaid whiche said latter booke tablets(?) begynynge
at the same tyme the former booke did ende viz:
begyninge upon the 27 daye of October 1604: it apearethe
oute of the partycular Cause & Sommes mencioned in the
said later booke that the said accomptant Willm Henshawe
did Receive for colles sould by hym since the said xxvii
day of October 1604 then lyeinge redy gotten & stacked
upon the banke(s) and also for . . .   & other coles theare gotten
at the same cole pitte since the said 27 of October 1604
(over and besyde the colles that weare then lyeinge upon the
banke Redye gotten) the said accomptant hath by
his owne confession so apearinge in his said
later booke receaved. viz

	(The Moneys wch this accomptant	)		)
	(Received apearinge in thys his	)		)
	(later booke for colles by sould	)		)
	(wch weare lyeinge Redy gotten	)		)
	(upon the bank the 27 daye	) l s d		)
	(of October 1604 as also moa(?) and)	68/7/11		)
	(other coales theare gotten since	)		)
	(betweene the said 27 of October	)		)
	(1604 and the 27 daye of	)		)
	(Aprill nexte followinge 1605	)		)
Receiptes in the	(			)
later booke	(And also more the said accomptant	)		) lxxxviii^l v^s ii^d

Receiptes in the later booke:

(The Moneys wch this accomptant   )
(Received apearinge in thys his   )
(later booke for colles by sould   )
(wch weare lyeinge Redy gotten   )
(upon the bank the 27 daye   ) l   s   d)
(of October 1604 as also moa(?) and) 68/7/11)
(other coales theare gotten since   )
(betweene the said 27 of October   )
(1604 and the 27 daye of   )
(Aprill nexte followinge 1605   )
(   )
(And also more the said accomptant   )   ) lxxxviii^l v^s ii^d
(Received in redye money from   )
(Sr Thomas Beawemont as in full satysfaction)
(of xl^l wch he was to paye for   )
(certien woodes (upon composition)   )
(formerly bought   ) l   s   d)
(of Mr Wyghteman the residue   ) 19/17/3)
(of the said 40^l beinge allowed   )
(& made even betweene the said   )
(Sr Tho Beawemonte, S^r   )
(Richard Leveson & ffr ffyton   )
(upon agreement to allowe an   )   whereof
(equall chance(?) for drawinge   )
(of water at the pytte wrought   )
(by sr Tho Beawemont as   )
(also by Sr Ric Leveson and ffr   )
(ffyton the some of   )

225

disbursement in | Whereof the same accomptant Willm Henshawe
the later booke | by this said later booke apeareth to have payed and
disbursed ffor the chalter colle of the said
coles newly gotten since the said xxvii day
of October 1604 untill the said 27 daye of
Aprill 1605 as by the partycularites thereof
in the said later booke maye apeare viz
ffor the chalter woorke $14/13^{s}/8^{d}$
And for the xtraordenery worke   )
of the same colles as maye   ) l s d
weekely by the partycularetees in) 14/13/6
the said booke apeare   )
And as so moche money by hym   )
disbursed the xv^{th} of Maye   )
for the behooffe of S^{r} Jo. Newdigate) l
$w^{c}h$ he then did leand to the   ) 25   ) $=lxvii^{l}xxii^{d}$
King upon a privie seale &   )   )
delyvered to thandes of Sr Richard)   )
Varney Knight the collector (?)   )   )
for the same   )   )
And also as so moche more delyvered(?))   )
by the said accomptant for the behoofe) l s d)
of the said Sr Jo Newdigate as in   )11/13/4)
the price of vii lyverey cloakes   )   )
boughte by ffr hollenshed at S^{r})   )
Jo Newdegates request the 22 of   )   )   Sic
June 1605   )   )   debett
And more for cariage of the said)s d   )
Cloakes from London to Coventree)4/8   )
by Kidney the carrier   )
And also more demanded by acct(?) & Ro Smythe
for makinge 20 hondred of ffyrsse $ffagottes_{e} w^{c}h$
weare employed aboute y^{e} colle woorke $16^{s}8^{d}$

So addinge together bouthe theis 2 severall totall summes of Receiptes viz
in the former booke $- Mxxv^{l}iiii^{s}xi^{d}$)
and in the later booke $- Lxxxviii^{l}v^{s}ii^{d}$) $MCxiii^{l}x^{s}i^{d}$
And lykewyse addinge together bouthe theis 2 severall
Tottalle Summes of disbursements   viz $xxi^{l}iii^{s}iiii^{d}$
In the former booke $- Mxxv^{l}iiii^{s}xi^{d}$) $Miiii^{xx}xii^{l}vi^{s}ix^{d}$
And in the later booke $- Lxvii^{l}xxii^{d}$)
Then the said accomptant Willm Henshawe nowe   Quos ----------(?)
apeareth upon this his accompte taken the xx^{th} of January   ixtu computante
1606 upon the perusall and castinge up of boathe his found xx^{mo} die Januarii
bookes as also his later booke to stand nowe & to be in debet this 1606 ad monis
xx^{th} of January 1606 hereupon the some of ________________ ffrancise
                                                                    ffyton

                                        Et sic Eque
                                          francis
                                           fyton